Industrial Lighting Systems

Industrial Lighting Systems

John P. Frier
and
Mary E. Gazley Frier

McGraw-Hill Book Company
New York St. Louis San Francisco Auckland
Bogotá Singapore Johannesburg London
Madrid Mexico Montreal New Delhi
Panama São Paulo Hamburg
Sydney Tokyo Paris
Toronto

Library of Congress Cataloging in Publication Data

Frier, John P

Industrial lighting systems.

Includes index.

1. Industrial buildings—Lighting. 2. Electric discharge lighting. I. Frier, Mary E. Gazley, joint author. II. Title.

TK4399.F2F74 621.32′25′4 79-23138

ISBN 0-07-022457-9

1234567890 KPKP 8987654321

The editors for this book were Tyler G. Hicks and Geraldine Fahey, the designer was Mark E. Safran, and the production supervisor was Paul Malchow. It was set in Baskerville by University Graphics.

Printed and bound by Kingsport Press.

Dedicated to
Jean, Kenneth, Catherine, Susan, and Ann

Contents

Preface

Despite substantial growth in the use of high-intensity discharge (HID) luminaires, very little has been written about how to use them. Every day, luminaire manufacturers receive requests for information on practical applications. There appears to be a definite need for this kind of material.

It was to meet this need that this book was written. It is a handbook for practical applications—to our knowledge, the only one ever written that deals exclusively with HID sources. These sources are commonly used in industry today because of their high efficacy. With the increased concern for energy conservation, this becomes a distinct factor in lighting design.

High-intensity discharge lamps are point sources, which means that they are adaptable to varied photometric control with reflectors of reasonable size. There is a wide choice of photometric distributions and lamp sizes, so that lighting systems can be designed to fit the needs of a particular situation.

This book can be used to aid in this design process or to evaluate a proposed lighting system. It should be of interest to anyone who designs lighting installations or must make a decision about a lighting system.

The book first gives background information, including an explanation of lighting terms and an analysis of the lighting equipment available. Then, for each type of lighting, information is given on how to read and understand luminaire photometric data. This is followed by step-by-step design procedures and sample problems. Included are graphs and charts that can be used to simplify the design process where several variables are involved.

Both indoor and outdoor design are covered, including the use of indoor industrial luminaires; roadway, area, and high-mast luminaires; and floodlights. The design information about floodlights is new material developed by the authors for this book. It is the most thorough discussion of outdoor lighting design that is available anywhere. Information is included on designing lighting for parking lots, roadways, and any other type of outdoor area. There is also an analysis of how to light an area to a minimum level of illumination.

Information is centered around an industrial facility, but the material is applicable to any area where HID luminaires are used.

The book is written in the metric system, with values in the standard United States system following in parentheses. All graphs and charts can be used with either system of measurement. This format will assist the lighting designer in the transition from one system to the other.

Some of the material in the book was developed by John P. Frier in his work for the General Electric Company and has been revised and used here with permission. The authors would like to thank the Lighting Systems Department of the General Electric Company for this cooperation.

The authors are also grateful to the following individuals for their personal help: Mr. John E. Kaufman, Technical Director, Illuminating Engineering Society of North America; Mr. Nels M. Andersen, Standards Publications Editor, National Electrical Manufacturers Association; Mr. Thomas F. O'Connor, CCTV Applications and Sales, Tube Products Department, General Electric Company; Mr. Ron Paugh, Manager Technical Services, High Intensity and Quartz Lamp Department, General Electric Company; and Mr. Joe Nuckolls, Manager Ballast and Control Engineering, Lighting Systems Department, General Electric Company.

John and Mary Ellen Frier

1

Lighting Terminology and Recommended Light Levels

In the past, the standard approach to lighting design was to recommend ever-increasing light levels. Light sources were not so efficient as they are today, and increased light levels often led to greater worker satisfaction and productivity.

Now, however, very efficient light sources are available, and there is a universally accepted need for energy conservation. Long before the present high costs of electricity, lighting designers were beginning to question how much light was enough. Research has been going on for years into how persons see, what influences their ability to perform a task, and other related questions. This research is continuing and is yielding answers, particularly for the performance of office tasks.

These studies have influenced all areas of lighting so that the fundamental approach to a lighting problem has changed. The approach now is to illuminate an area, not just to produce a certain amount of light. The expert lighting designer will look not just at the size of the area to be lighted but at what sort of activity will be performed in the space. If more than one task will be done in the area, then the space can be divided into smaller sections to aid in designing the lighting system. For instance, a room may be used only for storage at one end, but the other may contain machinery manufacturing the product that is to be stored. The two areas may require different approaches to lighting design. In addition, of course, no lighting designer will approach a problem without being very aware of energy costs. The power budget for the industrial site must be allocated to the various power needs. The lighting designer must recommend the most efficient, least costly lighting system that will be adequate for the work to be done.

Figure 1-1 shows an attractive, well-lighted work environment with luminaires placed at a very low mounting height. Such designs can only be accomplished with a thorough knowledge of lighting design procedures and the equipment available. New sources, new luminaires, and new ways of computing lighting needs all help the designer to choose the best possible lighting system. This in turn makes the space a pleasant place to work and aids in employee morale and performance.

The purpose of this book is to provide enough information for a designer to feel confident that a lighting system developed with these guidelines will be satisfactory. The book is intended as a practical guide to industrial lighting. Theoretical information is included only insofar as it aids in a practical approach to a lighting problem. Information is given for all areas of an industrial site, outdoors as well as indoors, but the general approaches to lighting design are applicable to other lighting situations as well.

High-intensity discharge (HID) sources—mercury, metal halide, and

Fig. 1-1 An attractive, well-designed lighting system using luminaires intended for low mounting heights.

high-pressure sodium (HPS)—have been dealt with almost exclusively. Incandescent and fluorescent sources have been in use many years, and design procedures for them are fairly well known. Little material that has practical application has been written for HID sources, however, and so the authors have concentrated on their use. These sources are highly efficient and fit in with today's emphasis on cost and energy savings.

The Language of Lighting

Those interested in lighting design are caught betwixt and between: the metric system is being phased in to general use in the United States, but the lighting practitioner is more familiar with the American system of feet and footcandles, and is reluctant to change.

The modern metric system is known as the International System of Units, and is usually referred to as the SI system, which stands for Système Internationale. There is a relationship in lighting design between the old and the new systems, which should be understood by anyone working in this field. Lighting is based on certain basic units of lighting measurement and certain fundamental laws.

BASIC UNITS OF LIGHT MEASUREMENT

The unit of luminous flux is the *lumen* in both systems. This is defined as the flux emitted within a unit solid angle (one steradian) from a uniform point source of one candela.

The *candela* is a unit of luminous intensity emitted by a light source in a given direction. It is defined as the luminous intensity of 1/600,000 square meter of a radiating cavity at the temperature of freezing platinum. The candlepower or luminous intensity of a particular lamp or luminaire is expressed as a certain number of candelas.

Luminous intensity must be expressed in terms of the direction of the incident or emitted radiation. Luminous intensity at various angles is portrayed graphically in the form of candela distribution curves for a particular lamp or luminaire. Directions for interpreting these curves are given in the lighting design chapters.

When luminous flux, or lumens, falls upon a surface, the surface is said to be *illuminated.* The effect is called *illumination,* and the unit for measuring illumination in the American system is the *footcandle.* A footcandle is numerically equal to an incident flux density of one lumen per square foot; that is, one lumen uniformly distributed over one square foot produces an illumination of one footcandle.

The SI term used in place of *footcandle* is *lux.* One lux is the illumination produced by one lumen uniformly distributed over one square meter.

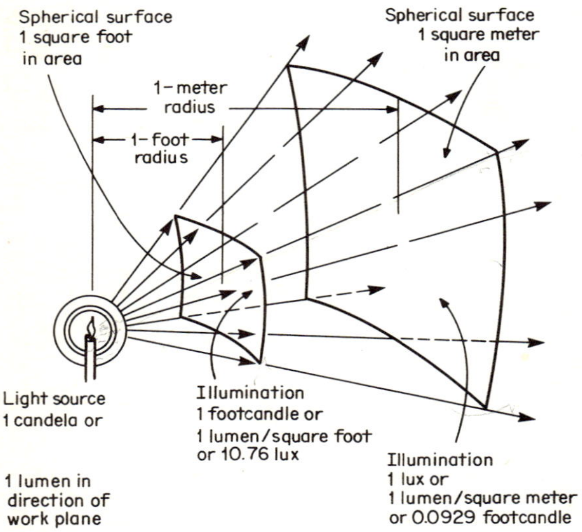

Fig. 1-2 Relation between footcandles and feet and lux and meters.

One lux equals 0.0929 footcandle, and one footcandle equals 10.76 lux. The relation between candelas, lux, and footcandles is shown in Fig. 1-2.

The authors have used both SI units and U.S. Customary units in all lighting calculations and recommendations. The SI units have been given first, with feet and footcandles following, usually in parentheses. In some cases, detailed charts expressed in footcandles have been left in that mode, with conversion factors given below.

Many lighting practitioners prefer to use *dekalux* (0.929 footcandle) rather than lux, since 1 dekalux is closer to the value of 1 footcandle. Lighting-level recommendations of the Illuminating Engineering Society of North America (IES) are given in dekalux.

Charts of lighting-level recommendations are rounded off to some reasonably close number. The authors have preferred to use actual SI equivalents to footcandles, since this ensures the integrity of the various formulas used in the discussion and sample problems. Since all numbers are given in both systems, it was more accurate to use only 10.76 lux as the equivalent of 1 footcandle. The authors have also stayed with the term *lux,* rather than use *dekalux* or *hectolux* (9.29 footcandles).

FUNDAMENTAL LAWS OF LIGHTING

The inverse-square law expresses mathematically the relationship between luminous intensity and illumination. It states that illumination at a point on a surface is directly proportional to the luminous intensity of the incident light at that point and inversely proportional to the square of its distance from the source. When the point is on a surface normal to the incident light, the following formula applies:

$$E = \frac{\text{cd}}{D^2}$$

where E = illumination (lux or footcandles)
cd = candelas directed toward the point of interest
D = distance from the light source to the point of interest

The law is based upon the concept of a point source which radiates equally in all directions. Under this condition the flux contained in a unit solid angle would spread over an increasingly larger area as the distance from the source increased.

Therefore the flux density, or lumens per square meter or square foot, decreases inversely as the square of the distance. This is illustrated in Fig. 1-3. At a 1-meter distance from the source of 1 candela, the illumination is 1 lux per square meter; at 2 meters, the illumination is ¼ lux; at 3 meters it is 1/9 lux; and so on. As the distance is doubled from the light source, the area covered by a given solid angle is quadrupled. Therefore the illumination is only one-fourth as great.

The inverse-square law applies when the surface to be lighted is perpendicular to the beam. When the beam of light strikes a surface at an angle the flux density, or lumens per square meter, on the inclined surface is reduced. The area intercepted by such a beam can be shown to be

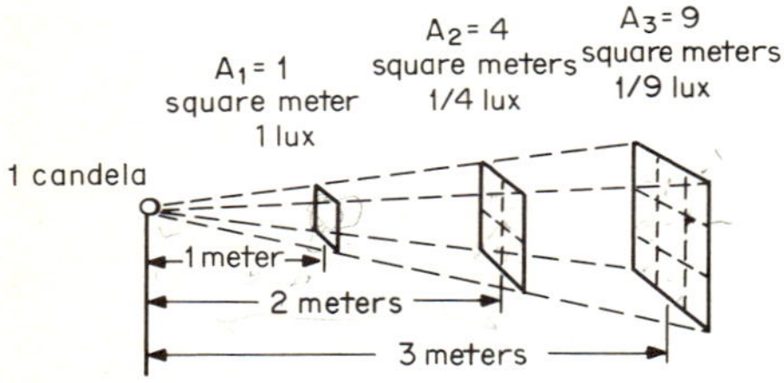

Fig. 1-3 Flux density, or lumens per square meter, decreases inversely as the square of the distance, as represented here.

proportional to the cosine of the angle which the inclined plane makes with the normal plane. The cosine law, or Lambert's cosine law, states that the illumination of a surface is proportional to the cosine of the angle of incidence of the light beam. When the light comes straight down to the surface, the angle of incidence is 0 degrees and the cosine is 1.0.

The inverse-square law and the cosine law combine to read

$$E = \frac{\text{cd} \cos \theta}{D^2}$$

where θ stands for the angle of incidence. This formula is used in many lighting calculations.

GLOSSARY OF TERMS

In addition to the terminology explained above, there are other terms commonly used in lighting design. A glossary of some of these words and phrases is included at the end of this chapter. This is a general reference only; refer to the *IES Lighting Handbook* for a much more complete dictionary. Many expressions will be explained in the text at the appropriate point.

Recommended Levels of Illumination

The levels of illumination recommended for industrial tasks by the IES are the American National Standards Institute (ANSI) standard for the

Table 1-1 Levels of illumination currently recommended by the Illuminating Engineering Society of North America

	Footcandles on Tasks*	Dekalux on Tasks*†
AIRCRAFT MANUFACTURING		
Stock parts		
Production	100	110
Inspection	200	220
Welding		
General illumination	50	54
Precision manual arc welding	1000‡	1080‡
Final assembly	100	110
ASSEMBLY		
Rough (easy seeing)	30	32
Rough (difficult seeing)	50	54
Medium	100	110
Fine	500‡	540‡
Extra fine	1000‡	1080‡
AUTOMOBILE MANUFACTURING		
Frame assembly	50	54
Chassis assembly line	100	110
Final assembly, inspection line	200	220
Body manufacturing		
Parts	70	75
Assembly	100	110
Finishing and inspection	200	220
BUILDING EXTERIORS		
Entrances		
Active	5	5.4
Inactive	1	1.1
CHEMICAL WORKS	30	32
COAL YARDS (protective)	0.2	0.2
ELECTRICAL EQUIPMENT MANUFACTURING		
Impregnating	50	54
Insulating: coil winding, testing	100	110
FOUNDRIES		
Annealing (furnaces), cleaning	30	32
Core making		
Fine	100	110
Medium	50	54
Inspection		
Fine	500‡	540‡
Medium	100	110
Molding		
Medium	100	110
Large	50	54
Pouring	50	54
INSPECTION		
Ordinary	50	54
Difficult	100	110
Highly difficult	200	220
Very difficult	500‡	540‡
Most difficult	1000‡	1080‡

Table 1-1 Levels of illumination currently recommended by the Illuminating Engineering Society of North America (*Continued*)

	Footcandles on Tasks*	Dekalux on Tasks*†
IRON AND STEEL MANUFACTURING		
Open hearth		
Stockyard, calcining building	10	11
Pouring slide		
Slag pits	20	22
Control platforms	30	32
Mixer building	30	32
Rolling mills		
Blooming, slabbing, hot strip, hot sheet, cold strip, plate	30	32
Pipe, rod, tube, wire drawing	50	54
Tinplate mills	50	54
Motor room, machine room	30	32
Inspection: black plate, bloom and billet chipping	100	110
LOADING AND UNLOADING PLATFORMS	20	22
Freight car interiors	10	11
MACHINE SHOPS		
Rough bench and machine work	50	54
Medium bench and machine work, ordinary automatic machines, rough grinding, medium buffing and polishing	100	110
Fine bench and machine work, fine automatic machines, medium grinding, fine buffing and polishing	500‡	540‡
Extrafine bench and machine work, grinding, fine work	1000‡	1080‡
MATERIALS HANDLING		
Wrapping, packing, labeling	50	54
Picking stock, classifying	30	32
Loading, trucking	20	22
Inside truck bodies and freight cars	10	11
PAINT SHOPS		
Dipping, simple spraying, firing	50	54
Fine hand painting and finishing	100	110
PAPER MANUFACTURING		
Beaters, grinding, calendering	30	32
Finishing, cutting, trimming, paper-making machines	50	54
Hand counting, wet end of paper machine	70	75
Paper machine reel, paper inspection, and laboratories	100	110
Rewinder	150	160
PARKING AREAS		
Self-service parking area	1	1.1
Attended parking area	2	2.2

	Footcandles on Tasks*	Dekalux on Tasks*†
RAILROAD YARDS		
Retarder classification yards		
Receiving yard		
Switch points	2	2.1
Body of yard	1	1.1
Hump and car rider classification yard		
Hump area	5	5.4
SHEET METAL WORKS		
Miscellaneous machines, ordinary bench work, presses, shears, stamps, spinning, medium bench work, punches	50	54
Tinplate inspection, galvanized; scribing	200	220
SHIPYARDS		
General	5	5.4
Ways	10	11
Fabrication areas	30	32
STORAGE ROOMS OR WAREHOUSES		
Inactive	5	5.4
Active		
Rough bulky	10	11
Medium	20	22
Fine	50	54
STRUCTURAL-STEEL FABRICATION	50	54
TESTING		
General	50	54
Extrafine instruments, scales, etc.	200‡	220‡
WELDING		
General illumination	50	54
Precision manual arc welding	1000‡	1080‡
WOODWORKING		
Rough sawing and bench work	30	32
Sizing, planing, rough sanding, medium-quality machine and bench work, gluing, veneering, cooperage	50	54
Fine bench and machine work, fine sanding and finishing	100	110

*Minimum on the task at any time for young adults with normal and better than 20/30 corrected vision.

†Dekalux is an SI unit equal to 0.929 footcandle: 1 dekalux = 10 lux.

‡Obtained with a combination of general lighting plus specialized supplementary lighting.

SOURCE: Reprinted with permission from *American National Standard Practice for Industrial Lighting*, A11.1-1973, RP-7, Illuminating Engineering Society of North America, New York, 1973.

industry. Some of these recommendations are included in Table 1-1. For a more complete list of lighting recommendations, refer to the *American National Standard Practice for Industrial Lighting,* A11.1-1973 RP-7.

The ANSI standards are the recommended minimum levels on the task at any time for young adults with normal and better than 20/30 corrected vision. Older workers, or those with poor eyesight, may require higher levels of illumination in order to see the designated task as well as young adults with normal vision.

These levels are recommended maintained values. Light output of any lighting system decreases over time not only because of lamp lumen depreciation (LLD) but also because of other light-loss factors such as dirt accumulation on luminaires and room surfaces. When the lighting system is designed, these factors should be taken into account and the system designed for higher initial light levels, so that the recommended levels will be maintained over time. Ways of doing this are explained in the appropriate design chapters.

Average horizontal levels in such areas as parking lots should take into account uniformity ratios and minimum values, as explained in the chapters on outdoor lighting.

Recommended lighting levels for specific visual tasks are measured on the work plane, whether it is horizontal, vertical, or in between. These levels can be met with a combination of general and supplementary or task-oriented lighting, or with general lighting alone.

The highest illumination levels have been designed for tasks requiring fine detail, low contrast, and prolonged work periods, such as fine bench work or detailed assembly tasks. Tasks that are less demanding have lower recommended values. Recommended levels are target values. Actual installations can be slightly higher or lower and still be adequate.

Design is only as accurate as the information used in making computations, and great care should therefore be taken to gather as much specific information as possible before designing a lighting system. Even so, measured light levels will seldom match calculated levels exactly. Manufacturing tolerances, slight errors in estimating light-loss factors or reflectances of surfaces, and other factors can change the actual, installed levels of illumination. In most cases, however, the design will still be satisfactory if recommended design procedures have been followed.

Throughout this book, levels of illumination have not been rounded off, but this is not intended to mean that slight variations in illumination level will make a difference in the quality of the lighting design. On the contrary, a change from 100 to 110 lux would never be noticed. Variations of 10 to 15 percent are well within the normal accuracy of present lighting design methods.

LEVELS FOR SAFETY

Light levels recommended for safety are much lower than those recommended for productivity. Table 1-2 presents the IES-recommended levels. These are the absolute minimum levels for safety of personnel at any time and at any location on any plane where safety is related to visual conditions. This lighting is intended to provide an awareness of potential hazards and the visibility to locate and avoid them.

Some situations may require a different approach. For instance, in areas where light-sensitive materials are manufactured, handled, or used, alternative methods of ensuring safe operations must be installed. The

Table 1-2 Levels of illumination for safety

Hazards Requiring Visual Detection	Slight		High	
Normal Activity Level	Low	High	Low	High
Illumination Levels				
Footcandles	0.5	1.0	2.0	5.0
Dekalux	0.54	1.1	2.2	5.4

SOURCE: Reprinted with permission from *American National Standard Practice for Industrial Lighting,* A11.1-1973, RP-7, Illuminating Engineering Society of North America, New York, 1973.

IES has also published guidelines for levels of illumination in specific industries, such as nuclear power plants and the petroleum and chemical industries. These should be referred to for more specific recommendations.

RECOMMENDED MAXIMUM LUMINANCE RATIOS

It takes a certain amount of time for eyes to adapt from one light level to another. If there are great differences in light levels within a worker's field of view, it will tire the worker's eyes to have to adapt continuously from one to the other. Also, the situation could be very uncomfortable and could reduce visual performance.

Table 1-3 gives the IES recommended maximum luminance ratios. If these recommendations are followed, the workplace will be more pleasant and workers will be able to perform tasks more comfortably and efficiently.

These are intended as maximum ratios. If possible, lower ratios are more desirable. This is particularly true for older workers, because older eyes lack the adaptability of younger eyes. Pupillary changes in the older eye lead to greater pupillary restriction and an inability to dilate readily. If luminance ratios are too high, discomfort can result.

Government Involvement

Government agencies are involved in the setting of recommended lighting levels and the setting of standards for energy conservation. The Occupational Safety and Health Administration (OSHA) of the Department of Labor, the Mining Enforcement and Safety Administration of the Department of the Interior, and the Department of Energy (DOE) have all issued regulations which affect the lighting industry. Other agencies and departments are also involved, on the federal and state levels.

OSHA

In general, most government lighting-level recommendations follow the IES recommendations, which have been approved by ANSI. However, there are other regulations as well. Some of the requirements issued by OSHA are included in Table 1-4.

Table 1-3 Recommended maximum luminance ratios

	Environmental Classification*		
	A	B	C
1. Between tasks and adjacent darker surroundings	3 : 1	3 : 1	5 : 1
2. Between tasks and adjacent lighter surroundings	1 : 3	1 : 3	1 : 5
3. Between tasks and more remote darker surfaces	10 : 1	20 : 1	†
4. Between tasks and more remote lighter surfaces	1 : 10	1 : 20	†
5. Between luminaires (or windows, skylights, etc.) and surfaces adjacent to them	20 : 1	†	†
6. Anywhere within normal field of view	40 : 1	†	†

*A = interior areas where reflectances of entire space can be controlled in line with recommendations for optimum seeing conditions; B = areas where reflectances of immediate work area can be controlled, but control of remote surround is limited; C = areas (indoor and outdoor) where it is completely impractical to control reflectances and difficult to alter environmental conditions.

†Luminance ratio control not practical.

SOURCE: Reprinted with permission from *American National Standard Practice for Industrial Lighting,* A11.1-1973, RP-7, Illuminating Engineering Society of North America, 1973.

Table 1-4 OSHA Regulations

General construction area lighting:	Minimum 5 footcandles
General construction areas, concrete placement, excavation and waste areas, accessways, active storage areas, loading platforms, refueling, and field maintenance areas:	Minimum 3 footcandles
Indoor: warehouses, corridors, hallways, and exitways:	Minimum 5 footcandles
Tunnels, shafts, and general underground work areas:	Minimum 5 footcandles; but minimum of 10 footcandles is required at tunnel and shaft heading during drilling, mucking, and scaling. Bureau of Mines–approved cap lights shall be acceptable for use in the tunnel heading.
General construction plant and shops (e.g., batch plants, screening plants, mechanical and electrical equipment rooms, carpenter shops, rigging lofts and active storerooms, mess halls, indoor toilets, and workrooms):	Minimum 10 footcandles
First-aid stations, infirmaries, and offices:	Minimum 30 footcandles
Manlifts:	a. 5 footcandles at each floor landing at all times when in operation. b. Both runs illuminated at all times to not less than 1 footcandle when lift in operation.
Powered industrial trucks:	a. In operating areas, controlled lighting of adequate intensity (ANSI levels). b. Where general lighting less than 2 lumens per square foot, auxiliary directional lighting required on the truck.
Pulp, paper, and paperboard mills:	a. Emergency lighting where necessary for employees to remain at machines or stations to shut down equipment in case of power failure; at stairways, passageways, or aisleways used by employees for emergency exit; at first-aid and medical facilities. b. For handling and storage of pulpwood and pulp chips, when loading or unloading after dark, ANSI standards. c. Permanent lighting in all areas where employees required to make machine adjustments and sheet transfers, according to ANSI standards.
Pulpwood harvesting:	Adequate lighting on roads when night work is necessary.

Table 1-4 OSHA Regulations (*Continued*)

Sawmills:	Adequate general and local lighting for rooms, buildings, work areas during time of use. Also adequate exits and lighting for fuel houses and bins.
Spray finishing using flammable and combustible materials:	a. When spraying areas illuminated through glass or other transparent material, only fixed lighting units are to be used. Panels must isolate the lighting from the spraying area. Panels must be arranged so that no residue will be raised in temperature to the danger point. b. Lamps outside spraying area to be totally enclosed and have guards. c. Any portable lamps to be Class I Hazardous type.
Laundry machinery and operations:	Lighting to meet ANSI standards for industrial lighting.
Overhead and gantry cranes:	Light in cab sufficient for operator to see clearly.
Textiles:	Lighting to ANSI standards.
Means of egress:	a. Adequate and reliable illumination provided for exit facilities. b. Exit signs illuminated to 5 footcandles on surface.

MINING

The Mining Enforcement and Safety Administration has issued regulations for underground coal mines as well as surface coal mines and surface work areas of underground coal mines. The surface rules include not only average illumination intensity levels but also uniformity ratios. The average illumination intensities are from 0.5 to 10 footcandles. Uniformity ratios are either 5:1 or 10:1. Designing for a certain uniformity requires sophistication in lighting design; techniques described in this book should be followed exactly in order to ensure satisfactory results.

DOE

The Department of Energy absorbed the functions of the Federal Energy Administration (FEA) in late 1977. At the time of this writing, the DOE is in the process of developing a document entitled *Buildings Energy Performance Standards.* These standards will cover overall design budgets for new buildings and will give guidelines which will aid in energy conservation.

In addition, the federal government is providing grants to state governments to encourage them to establish energy conservation programs.

There are two state grant programs. In order to qualify, states must establish lighting efficiency standards (among others) for new or existing public buildings. This involves setting energy budgets and staying within the prescribed limits. Some states are applying these requirements to residential and nonresidential buildings as well as public buildings.

The energy conservation guidelines given by the American Society of Heating, Refrigerating, and Air Conditioning Engineers Inc. (ASHRAE) in *Energy Conservation in New Building Design,* ASHRAE Standard 90-75, are being used as a guide. This document is being revised at the time of this writing, under the sponsorship of ASHRAE and IES. More detailed information on this standard is included in Chap. 16, Lighting Economics. Chapter 16 also discusses the ASHRAE standard *Energy Conservation Existing Facilities—Industrial,* 100.4P, IES EMS 4.4, which is cosponsored by ASHRAE and IES and is in its second draft at the time of this writing.

The lighting designer should obtain copies of these documents, as well as any others that apply, in order to ensure that the lighting system being designed will meet all federal and state guidelines.

Glossary of Lighting Terms

Asymmetric Distribution: A light distribution in which the curves of vertical distribution are not the same for all planes.

Ballast: A device used with an electric-discharge lamp to obtain the necessary circuit conditions for starting and operating.

Beam Spread: The angular divergence of a beam, measured in degrees.

Bracket (Mast Arm): An attachment to a lamp post or pole from which a luminaire or lighting fixture is suspended.

Candela: The unit of luminous intensity. One candela is defined as the luminous intensity of 1/600,000 square meter of projected area of a laboratory device called a blackbody radiator which is operated at the temperature of solidification of platinum.

Candlepower: Luminous intensity expressed in candelas. It is the amount of light that will illuminate a surface one meter distant from the light source to an intensity of one lux, or one foot distant from the light source to an intensity of one footcandle.

Candlepower Distribution Curve: The intensity of luminous radiation expressed in candlepower, measured at various angles about a light source, and graphically represented.

Coefficient of Utilization (CU): The ratio of useful light on the lighted surface to the total emitted by the lamp in the luminaire, i.e., ratio of effective lumens to bare-lamp lumens.

Efficacy: The quotient of the total luminous flux emitted by a light source to the total power input to the source. In the case of an electric lamp, efficacy is expressed in lumens per watt.

Electric-Filament Lamp: A light source consisting of a glass bulb containing a filament electrically maintained at incandescence; commonly called an incandescent lamp, filament lamp, incandescent filament lamp, electric light, or light bulb.
Electric-Discharge Lamp: A lamp in which light is produced by the passage of electricity through a metallic vapor or gas, such as mercury, sodium, neon, argon, and other metallic additives, etc., enclosed in a tube or bulb; sometimes called a vapor lamp.
Flux: The time rate of flow of light.
Footcandle: A unit of illumination. It is defined as the illumination on a surface one square foot in area on which a flux of one lumen is uniformly distributed. One footcandle is equal to a luminous intensity of one lumen per square foot.
Footlambert: A unit of photometric brightness, equal to the average brightness of any surface, or the uniform brightness of a perfectly diffusing surface, emitting or reflecting one lumen per square foot.
Glare: The sensation produced by brightnesses within the visual field that are sufficiently greater than the luminance to which the eyes are adapted to cause annoyance, discomfort, or loss in visual performance and visibility.
Direct Glare: Glare resulting from high brightnesses or insufficiently shielded light sources in the field of view.
Disability Glare: Glare which reduces visual performance and visibility and which is often accompanied by discomfort.
Discomfort Glare: Glare which produces discomfort. It does not necessarily interfere with visual performance or visibility.
Reflected Glare: Glare resulting from specular reflections of high brightnesses in polished or glossy surfaces in the field of view. It is usually associated with reflections from within a visual task or areas in close proximity to the region being viewed.
Hazardous Location: An area where ignitable vapors, dust, fibers, or flyings may cause a fire or explosion created by energy emitted from lighting or other equipment or by electrostatic generation.
High-Intensity Discharge (HID) Lamps: Mercury, metal halide, and high-pressure sodium (HPS) lamps.
Illumination: The density of luminous flux incident on a surface.
Isocandela Line: A line plotted on any appropriate coordinates to show directions in space about a source of light in which the candlepower is the same.
Isoillumination Curve: A curve plotted on any appropriate coordinates to show points of equal illumination, expressed in lux or footcandles.
Lamp Lumen Depreciation (LLD) Factor: The multiplier to be used in illumination calculations to relate the initial rated output of light sources to the anticipated minimum rated output based on the relamping program to be used.
Light Center Length: The distance from some reference point on the base, generally the bottom contact of an incandescent filament lamp, to the center of the filament. In mercury-vapor or sodium lamps the light center length is the distance from the bottom contact of the lamp to the center of the arc.
Light-Loss Factors: Lighting maintenance factors made up of many components. These are the most important:

1. Temperature, voltage, and ballast performance
2. Aging of luminaire finish and material
3. Accumulation of dirt on interior surfaces
4. Burned-out, unreplaced lamps
5. Lamp lumen depreciation
6. Luminaire dirt depreciation

Lumen: A unit of luminous flux that equals the flux emitted through a unit solid angle (one steradian) from a uniform point source of one candela; unit in which the output of a light source is frequently expressed.
Luminaire: A complete lighting device consisting of a light source together with other parts such as a globe, reflector, refractor, housing, and such support as is integral to the housing. The pole, post, or bracket is not considered a part of the luminaire.
Luminaire Dirt Depreciation (LDD) Factor: The multiplier to be used in illumination calculations to relate the initial illumination provided by clean, new luminaires to the reduced illumination that they will provide because of dirt collection on the luminaires at the time at which it is anticipated that cleaning procedures will be instituted.
Lux: The International System (SI) unit of illumination. One lux is equal to a luminous intensity of one lumen per square meter. 1 lux = 0.0929 footcandle.
Mounting Height (MH), Indoor Industrial: The vertical distance from the work plane to the light center of the luminaire, or to the plane of the ceiling, for recessed equipment.
Mounting Height (MH), Roadway: The vertical distance between the roadway surface and the center of the apparent light source in the luminaire.
Optical Assembly: That part of the luminaire which consists of the light-controlling elements such as the globe, refractor, and reflector.
Overhang: The distance between a vertical line passing through the luminaire and the curb or edge of the street or highway.
Reflectance: The ratio of the light reflected by a surface to the light incident upon it.
Reflector: A device used to redirect the luminous flux from a source by the process of reflection.
Refractor: A glass or plastic band, bowl, or globe designed to control the direction of the emergent light rays by refracting prisms.
Silhouette: The effect produced when an object is seen as a dark outline against a brighter background. This condition occurs in street lighting when the dominant light source is beyond the object viewed.
Spacing: The distance in meters or feet between successive lighting units, measured along the centerline of placement. On a roadway, the distance would be measured along the center of the street.
Stray Light: Light falling outside the desired beam. This light is wasted by a luminaire or optical system and often produces an undesirable lighting effect.

Symmetrical Light Distribution: A light distribution in which the curves of vertical distribution are substantially the same for all planes.
Veiling Reflections: Regular reflections superimposed upon diffuse reflections from an object that partially or totally obscures the details to be seen by reducing the contrast.
Work Plane: In industrial lighting, generally considered to be a plane 0.76 to 0.9 meter ($2\frac{1}{2}$ to 3 feet) above the floor.

Bibliography

American National Standard Practice for Industrial Lighting, A11.1-1973, RP-7, Illuminating Engineering Society of North America, New York, 1973.

Barbrow, L. E.: "The Metric System in Illuminating Engineering," *Illuminating Engineering,* November 1967, pp. 638–640.

"Dictionary of Lighting Terms," *IES Lighting Handbook,* 5th ed., sec. 1, Illuminating Engineering Society of North America, New York, 1972.

Energy Conservation Existing Facilities—Industrial, ASHRAE Standard 100.4P, IES EMS 4.4, second draft, American Society of Heating, Refrigerating, and Air-Conditioning Engineers, Inc., and Illuminating Engineering Society of North America, New York, 1978.

Energy Conservation in New Building Design, ASHRAE Standard 90-75, American Society of Heating, Refrigerating, and Air-Conditioning Engineers, Inc., New York, 1975.

Flynn, John E., and Gregory J. Subisak: "A Procedure for Qualitative Study of Light Level Variations and System Performance," paper presented at the 1978 Annual National Technical Conference, Illuminating Engineering Society of North America, Denver, Colo., Aug. 27–31, 1978.

Glossary of Terms Used in Street and Highway Lighting, OLP-1227A, General Electric Company, Lighting Systems Business Department, Hendersonville, N.C., 1974.

Illumination, Occupational Safety and Health Reporter, Reference File, 31:3004, p. 8.

Kaufman, J. E.: "Introducing SI Units," *Illuminating Engineering,* October 1968, pp. 537–540.

Kaufman, John E.: "Lighting for Safety," *Lighting Design and Application,* March 1973, pp. 6–8.

Kaufman, John E.: "Must the Most Used Be the Most Abused?" *Lighting Design and Application,* November 1976, p. 56.

Lighting Photometric Data and Sample Calculations, GET-3100F, General Electric Company, Lighting Systems Department, Hendersonville, N.C., 1971.

Rowe, Gordon D.: "Understanding the New Language of Lighting," *Plant Engineering,* Jan. 21, 1971, pp. 68–69.

"Selected General Industry Safety and Health Standards," Department of Labor, Occupational Safety and Health Administration, *Federal Register,* Dec. 13, 1977.

Serchuk, Alan: "The Brighter Side of Lighting," *Factory,* March 1976, pp. 41–53.

"Surface Illumination Standards for Surface Coal Mines and Surface Work Areas of Underground Coal Mines, Proposed Rules," *Federal Register,* vol. 42, no. 9, Jan. 13, 1977, pp. 2805–2807.

"Underground Coal Mines, Illumination Systems; Promulgation Date, and Testing and Evaluation Procedures," part II, *Federal Register,* vol. 41, no. 64, Apr. 1, 1976, pp. 14102–14110 (Proposed Rules, Department of the Interior, Mining Enforcement and Safety Administration).

2

Lamps and Ballasts

High-intensity discharge (HID) lamps commonly used in industrial lighting include mercury, metal halide, and high-pressure sodium. The Illumination Engineering Society of North America (IES) characterizes these sources in the *IES Lighting Handbook* as discharge devices which are wall-stabilized and whose light-producing envelopes have a bulb wall loading in excess of 3 watts per square centimeter.

High-intensity discharge sources are noted for long life and high efficacy. They are compact sources, which give great flexibility in luminaire design.

Lamp and ballast manufacturers publish detailed information about their products. The purpose of this chapter is to give general information needed before choosing a lamp-ballast combination. The first half of the chapter deals with lamps; the second half dicusses ballasts.

Kinds of Sources

The three common kinds of HID sources are mercury, metal halide, and high-pressure sodium. All contain a sealed arc tube, and the light emitted is much like lightning in a bottle, except that the radiation is sustained rather than dissipated immediately. Power connected to the two ends of the arc tube excites a starting gas, which in turn heats a metal or metallic compound to a point where it ionizes and becomes the principal source of light. Each of the three types of source contains a different metal or combination of metals within the arc tube.

The energy given up in the form of radiation is of a specific wavelength in each type of lamp. As a result, the apparent color of each source is different, depending on whether mercury, metal halides, or sodium is the main metal used in establishing the arc. Anyone who has ever thrown into a fire a handful of powder containing various elements is familiar with this effect: some of the flame turns to blue, other parts to red or yellow, depending on which element is being excited.

Every HID source requires a ballast to limit current across the arc once the arc has been struck. If a ballast were not used, the lamp would destroy itself. The dynamic characteristics of each type of lamp are somewhat different, so that each, except for very special cases, requires its own specific type of ballast.

MERCURY

Mercury lamps were the first of the presently used general-lighting HID sources developed. They were noted for long life and low initial cost per footcandle when compared to incandescent lamps. Mercury lamps have been used where burning hours are long and system maintenance is difficult or expensive. These sources are now available from 40 to 1000 watts, in many shapes, to suit a range of industrial applications.

Figure 2-1 shows a typical mercury lamp, which consists of an outer bulb filled with an inert gas (usually nitrogen) which forms a protective layer around the inner arc tube. The arc tube contains high-purity mercury. Since mercury has a low vapor pressure at room temperature, argon gas is also present to ease starting.

When the lighting circuit is energized, the starting voltage is impressed across the gap between the main electrode and the starting electrode. This creates an argon arc that creates heat to vaporize the mercury. The ionized mercury particles reduce the resistance between the main electrodes to a point where the ballast causes the main arc to strike. When all the mercury within the arc tube has become vaporized, the lamp is at its *steady-state condition.* Since the resistance of the main arc is much lower that that of the starting circuit because of the starting resistor, the starting arc ceases. Warm-up to 80 percent of light output can take from 3 to 5 minutes. More time may be needed in extreme cold.

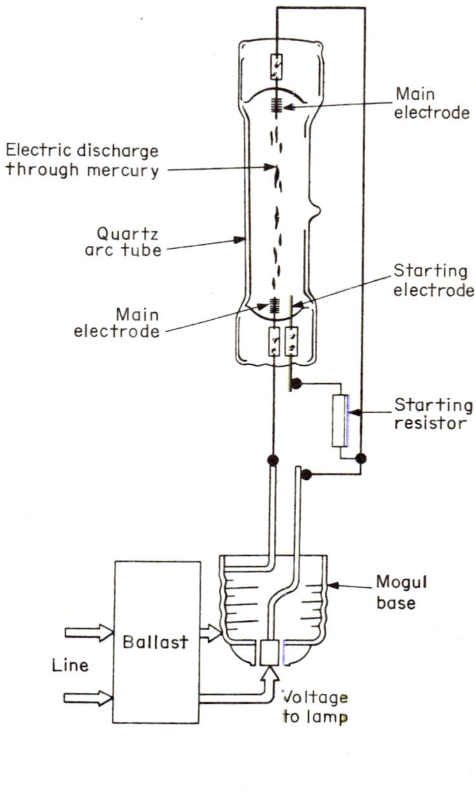

Fig. 2-1 Design of a mercury lamp. The inner arc tube is protected by an outer bulb filled with an inert gas, such as nitrogen.

The outer glass bulbs of mercury lamps either are clear glass or have phosphor coating on their inner surfaces. These types are electrically interchangeable if they are of the same wattage, but the photometric distribution of the luminaire will be different. The phosphor converts ultraviolet energy into visible light in the red portion of the electromagnetic spectrum and provides a better balance of light in terms of color.

One type of mercury lamp is known as *self-ballasted.* This means that the separate ballast necessary with all HID lamps has been replaced with a ballast circuit which consists of a filament within the lamp itself. This lamp can be used to replace incandescent lamps directly if the sockets are porcelain and the wiring adequate. Enclosed luminaires should be designed for at least the rated lamp wattage.

The main advantage of self-ballasted lamps over the incandescent lamp system is increased life. Self-ballasted lamps are used where high temperatures could damage external ballasts or where maintenance costs for replacing incandescent lamps are high.

Self-ballasted mercury lamps either can be conventional or can incorporate a solid-state circuit. The conventional self-ballasted lamp consumes

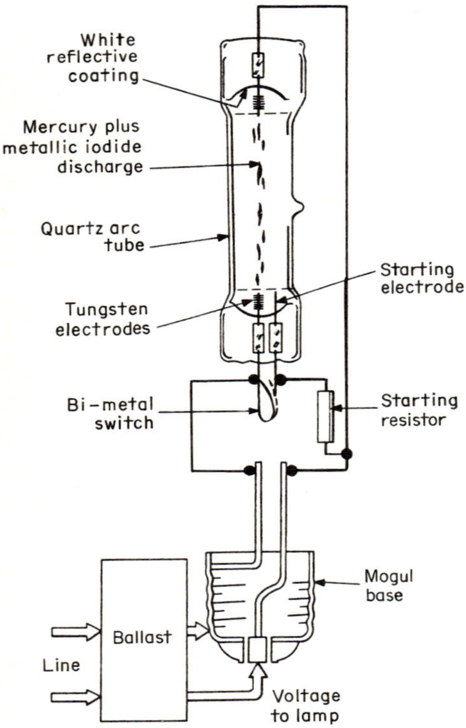

Fig. 2-2 Electrical details of a metal halide lamp. Metal halide lamps are similar to mercury lamps, except that metallic iodides are included with mercury in the arc tube.

about 60 percent of the wattage in the filament and 40 percent in the arc. A solid-state circuit can reverse these percentages. By comparison, the ballast in a normal lamp-ballast combination consumes 20 percent or less of the total wattage.

METAL HALIDE

Metal halide lamps are also known as *metallic vapor* or *metallic halide* lamps. They are available in sizes of 175 to 1500 watts. Metal halide lamps provide higher efficacy than mercury lamps and better color rendering.

The metal halide lamp is similar in design to the mercury lamp (see Fig. 2-2). The main difference is that other metals have been added to the mercury in the arc tube in order to improve radiation color characteristics and efficacy. Mercury is present in the arc tube, and other metals are added in the form of their halide salts, usually iodides. This creates an amalgam of mercury with these iodides. Some lamps contain two iodides in addition to the mercury, while others contain three.

As in mercury lamps, argon is used in metal halide lamps to strike the first arc and generate the initial heat. Gradually, as the temperature in the arc tube rises, the main arc is struck, and the various iodides vaporize and separate into elementary iodine and the two or three additive metals. Warm-up can take from 5 to 7 minutes, depending on lamp and ballast type.

The amounts of these materials that are present in the arc tube exceed what is needed to stabilize the arc. This creates a pool of excess amalgam which can be drawn upon when needed. The temperature of this pool of additives determines the pressure within the arc tube and the amount of additive in vapor form. This determines the chromaticity of the lamp as well as its light, or lumen, output. This means that variations in wattage due to system variations, lamp variations, and burning position will result in changes in apparent color.

Thus, for the same variation in wattage, metal halide lamps have a wider variation in color from lamp to lamp than mercury lamps. For those applications where lamp-to-lamp color consistency is of extreme importance, manufacturers should be consulted about the choice of lamp type.

Metal halide lamps are also available with phosphor-coated outer bulbs. This improves their color rendition slightly and reduces brightness. This is of particular importance for indoor lighting applications at low mounting heights.

HIGH-PRESSURE SODIUM

High-pressure sodium (HPS) sources have the highest efficacy of any white HID sources; that is, they provide more lumens per watt. As power rates have increased, so has the use of this lamp. High-pressure sodium lamps have about 50 percent higher efficacy than either metal halide or fluorescent sources, over twice the efficacy of mercury, and six times the efficacy of incandescent sources, as shown in Fig. 2-3.

High-pressure sodium lamps are different from other HID sources (see Fig. 2-4). Rather than a starting electrode, special starting circuitry in the ballast is used to start the lamp. The arc tube is made of alumina and uses xenon as a starting gas and an amalgam of mercury with sodium for the main arc. The ballast contains a special starting circuit which ionizes the xenon by means of low-energy, high-voltage pulses (about 2500 or 3000 volts). These pulses are superimposed on the 60-hertz voltage until the xenon gas is ionized enough for the open-circuit voltage (OCV) of the ballast to establish and maintain the xenon arc. After the xenon arc is struck, the starting pulses normally stop. More and more sodium is vaporized into the arc, resulting in color changes during the warm-up cycle of 3 to 4 minutes. However, the color of the lamp in its steady-state condition will not change appreciably over the life of the lamp and will change only slightly with changes in operating wattage.

The vapor pressure of the sodium is approximately 55 torr in order to obtain the desired efficiency and color. This corresponds to a 1000°C arc-tube–wall temperature. The arc tube is enclosed in a vacuum for heat conservation, and the outer envelope of the lamp is heat-resistant borosilicate glass, which is typically used as the outer-bulb material for most HID lamps.

Arc tubes in HPS lamps are made of polycrystalline alumina (PCA),

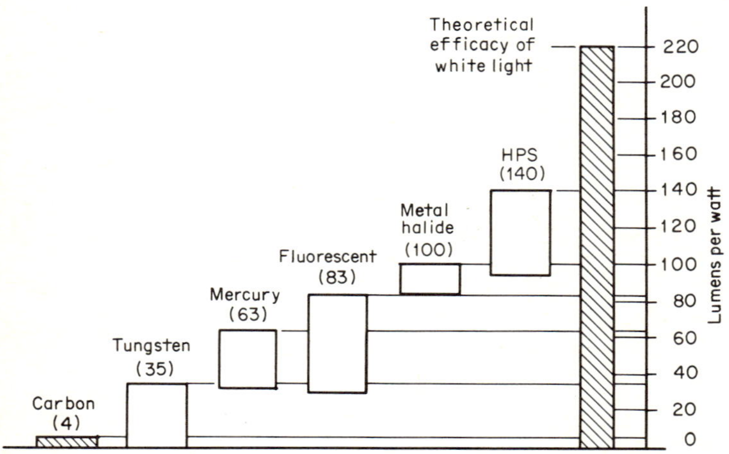

Fig. 2-3 Light-source efficacy in lumens per watt. High-pressure sodium sources have the highest efficacy of any lamps commonly used in industrial lighting.

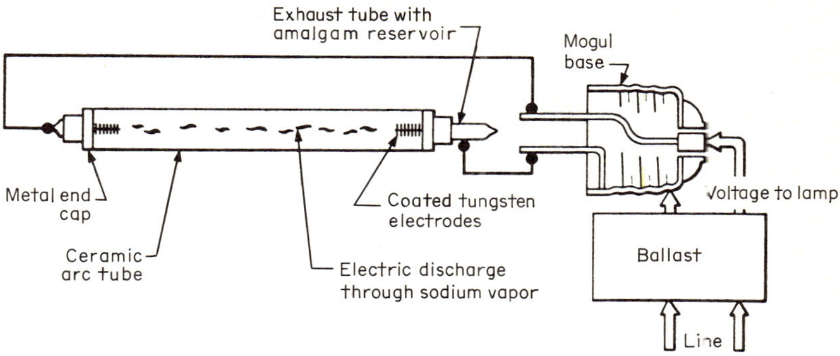

Fig. 2-4 Design of an HPS lamp. High-pressure sodium lamps require a special starting circuit located in the ballast. This circuit delivers a low-energy, high-voltage pulse of 2500 or 3000 watts.

which is translucent and resistant to alkali attack, and has a higher melting point than quartz.

Lamp Life and Lumen Maintenance

All lamps are given a rated life, expressed in burning hours. Table 2-1 gives the rated life and initial lumen output for some mercury, metal halide, and HPS sources. These values change frequently as lamps are improved; always check with manufacturers for the latest data. The figures for rated life are based on 10 hours per start. Initial lamp lumen ratings are measured after lamps have been burned for 100 hours.

Mercury lamps have a very long operating life—in most cases over 24,000 hours. Lamp life is defined as the point where 50 percent of the lamps have failed. For mercury lamps there are still 66 percent in operation at 24,000 hours, but since the lumen output at that point is very low, 24,000+ has been designated as the economic life of the lamp. Lumen maintenance, the light output throughout the life of the lamp, is so low after 24,000 hours that the mercury lamp provides light at the efficacy of an incandescent lamp. Thus, the life of the lamp is essentially over.

In HPS sources, lamp voltage rise is a determining factor in lamp life. As these lamps age, the lamp volts rise until the ballast can no longer sustain them. This is the normal end of life, at which point the lamp begins to cycle on and off.

There is evidence that HPS sources have improved life when operated continuously or when turned off and on infrequently. The exact reasons for this are related to the fact that lamp seals and electrodes are stressed as the lamp warms up. Thus, if lamps are warmed up less frequently, there is less stressing. Since normal lamp life is 24,000 hours, it is difficult to obtain test data to verify the actual increase in life due to continuous burning.

This increase is accompanied by improved lumen maintenance. The value of a lighting system is dependent not on lamp life but on the level of illumination the system provides. If the level of illumination drops significantly, the luminaires should be cleaned and relamped.

Table 2-1 Lamp performance—clear sources

Lamp Type	Wattage	Hours of Life*	Initial Lumens
High-pressure sodium			
	50	24,000	3,300
	70	24,000	5,800
	100	24,000	9,500
	150	24,000	16,000
	200	24,000	22,000
	250	24,000	27,500
	310	24,000	37,000
	400	24,000	50,000
	1000	24,000	140,000
Mercury			
	100	24,000+	4,200
	175	24,000+	8,600
	250	24,000+	12,100
	400	24,000+	22,500
	1000	24,000+	63,000
Metal halide			
	175	7,500	14,000
	250	10,000	20,500
	400	10,000	34,000
	1000	10,000	110,000
	1500	3,000	155,000

*At 10 or more hours per start for high-pressure sodium and metal halide, except 1500-watt, which is 5 hours per start.

Since the initial lighting system is designed to compensate for the lumen maintenance of the lamp, a shorter economic lamp life may be justified. For mercury lamps this is usually under 24,000 hours.

The system wattage will be dependent on the initial luminaire quantity and wattage. If a low maintenance is selected either due to the lamp or luminaire, the initial system may consume more power than necessary. A higher maintenance or light loss factor can be obtained by replacing lamps before they fully deteriorate. The break-even point occurs where the saving in power and initial system costs compensates for the higher lamp and maintenance costs.

Burning Position

High-pressure sodium sources are not affected by their burning position. Both mercury and metal halide lamps deliver their highest lumen output when operated vertically. Lamp life is also affected when a lamp is burned in a position other than vertical, and lumen depreciation is substantially higher.

These differences in lumen output are a result of thermal changes within the arc tube which cause it to bow upward when the lamp operates horizontally. Manufacturers often specify base-up or base-down operation for certain lamps. Some arc tubes have been specifically designed to be operated horizontally, but lamp efficacy may be reduced.

Color Rendition and Apparent Color Temperature

Metal halide lamps provide better color rendition than either mercury or HPS. However, if very high color rendering is required, such as in mixing paints or matching close color tolerances on a production line, supplementary fluorescent or other sources with high color-rendering indexes should be used.

There is usually little advantage to mixing different HID sources to improve color rendering. All are considered "white" light, and all have proved acceptable in industrial environments.

Over the life of clear mercury lamps, there is no noticeable change in chromaticity. Phosphor-coated mercury lamps, however, gradually become cooler in color because the phosphor slowly wears out. Apparent or correlated color temperature for new mercury lamps is 3300 or 3900 kelvins, depending on lamp type.

The apparent color temperature of metal halide lamps becomes lower over life, resulting in a shift toward warmer color. The apparent color temperature of clear metal halide sources varies from 3600 to 4250 kelvins. Phosphor-coated lamps are rated at 3400 to 3900 kelvins.

High-pressure sodium lamps will not change much in apparent color temperature during life. Generally, their light will appear whiter, and then go back to warmer color near the end of the lamps' life. Apparent color temperature is 1900 or 2100 kelvins for standard HPS sources, and 1800 kelvins for Penning start lamps.

For comparison purposes, note that the apparent color temperature of incandescent sources is 2890 kelvins.

Restriking

High-intensity discharge sources will not restart immediately if they are extinguished for any reason. It takes more voltage than the ballast can supply to re-ionize gases while they are hot and under pressure. In mercury or metal halide lamps, it can take several minutes for the gas to cool and for internal pressure to drop enough for the arc to be restruck.

High-pressure sodium sources can restrike in 1 minute, because the lamp does not have to cool as much as other HID lamps before the arc can be restruck. High-pressure sodium lamps which operate on mercury ballasts can have a restrike time of 3 minutes.

Lamp Designations

All HID lamps are assigned certain designations so that similar lamps of different manufacturers can have the same designations. The American National Standards Institute (ANSI) has developed the code shown in Fig. 2-5. In addition, some manufacturers have their own designation systems, which they use together with the ANSI code.

Effects of Ambient Temperature

Normally, mercury lamps are not materially affected by ambient temperature. However, very high temperatures, such as those above 210°C on all screw-base lamps, or above 400°C on the bulb wall, can lower perfor-

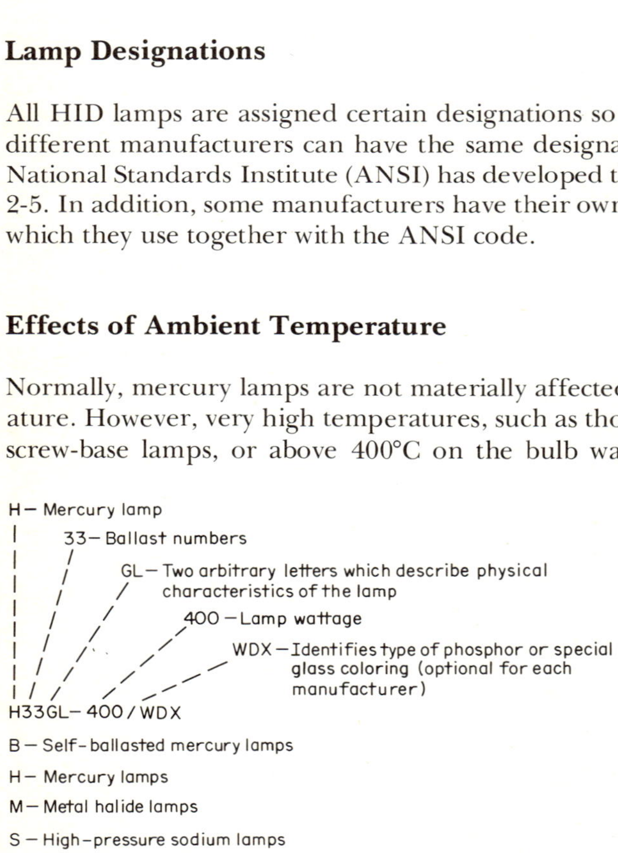

Fig. 2-5 The American National Standards Institute recommends this labeling system for HID sources. Manufacturers sometimes use a system of their own in addition to the ANSI system.

mance. Low ambient temperatures (below 20° F) could produce a condition where any HID lamp will not start at normal line voltage or will not fully warm up. This is mainly a result of ballast limitations; manufacturers published information should therefore be consulted.

Stroboscopic Effect

All HID sources vary in their light output when used on alternating currents. This variation is called *flicker.* When a flickering lamp is used to light a moving object, the object appears to flicker also. This is known as *stroboscopic effect.*

In industry, stroboscopic effect is most noticeable on rotating or oscillating machinery. It is almost impossible to eliminate when HID sources are used, but it usually is only a minor annoyance. Even if two lamps on different phases are mounted side by side, the image reflected from metal surfaces will be single-phase and may appear to flicker. This is particularly true at low mounting heights, where one luminaire usually contributes a major portion of the illumination to any one spot.

Stroboscopic effect is present with all HID sources, but it is particularly noticeable with HPS. Metal halide lamps operated on single-phase power are adequate for most sporting events and for many industrial applications. Mercury lamps can be used in pairs with lead lag ballasts; this cuts down or eliminates any noticeable flicker.

All sources can be operated on three-phase power to help overcome stroboscopic effect. Some HID sources can be used with direct current in special situations, such as when mounted on cranes. In that case, flicker is not a problem.

Lamp Failures

The outer envelope of HID sources is made of a glass that is highly resistant to attack by most environments. However, the screw shell and socket could be subject to environmental attack and must be covered by luminaires which have been specifically designed for the situation. Some manufacturers provide luminaires designed to be used in various types of hazardous locations. More information on this subject is given in Chap. 3, Luminaires and Their Functions.

Many sockets have a dimple in the screw thread which makes it difficult to screw the lamp into the socket. If the lamp is not screwed securely into the socket, the electric contact may not be good. This can cause arcing in the lamp base and socket, resulting in failed units.

Sockets are purposely made so that it is difficult to screw in the lamp, to prevent the lamp from shaking loose or backing out of the socket. But this feature makes it difficult to get sufficient tension if the lamps are changed with a pole changer from the floor. Poor electric contact will result unless the lamp changer has a sufficient grip on the lamp. The lamp should be turned into the socket to a torque of 35 inch-pounds.

Some HID lamps can be extinguished if subjected to extreme shock, although this probably won't take place unless the lamp is mounted on a moving crane.

If the outer jacket is broken off a mercury lamp, the lamp will not fail unless the arc tube cracks or the lead-in wires open up. Bare mercury arc tubes emit very short wave-length ultraviolet light, which can cause suntanning, conjunctivitis, or other skin and eye problems. If the outer bulb on a mercury lamp breaks off, the lamp should be turned off and allowed to cool, and then removed from the socket.

Some types of metal halide lamps need to be placed in enclosed luminaires because the arc tube may shatter at the end of life. This can break the outer bulb. Manufacturers should be consulted on this point.

Interchangeability of HID Sources

In most cases one HID lamp cannot be successfully interchanged with another that is of a different type or wattage. High-intensity discharge lamps have different starting voltages, different operating voltages, different operating currents, and different current waveshape requirements. Operating a lamp of a different type on a ballast—even if the lamp would start—would cause it to operate at other than its normal condition. This can shorten lamp life and possibly damage the ballast. In addition, this sort of substitution can void a manufacturer's warranty.

The only lamps that can be interchanged are those that have similar electrical characteristics and have been specially designed for such use. There are a number of lamp types which have been developed specifically for operation in existing systems: argon–metal halide lamps which will operate on certain types of mercury ballasts, certain mercury lamps which can operate in luminaires without a ballast (self-ballasted), and a number of HPS lamps which can operate on existing mercury systems. In all cases, care must be taken that the manufacturer's instructions are followed exactly.

These special lamps are used mainly as replacement lamps, because in most cases, over the long run, these lamps do not perform as well as lamps of the same wattage operated on standard ballasts designed specifically for them. In many cases the performance of the lamp is exaggerated, showing initial benefits over the lamp it replaces. However, a new system including a new ballast and lamp will almost always produce better long-term results than replacing only part of a system. Sometimes it is not practical to install an entirely new system; in such cases these special lamps can be used to upgrade an existing system and increase its efficiency.

High-intensity discharge lamps will operate on higher-wattage ballasts of the correct type. However, this will result in shortened life, poorer lumen maintenance, possible devitrification of the arc-tube quartz, and perhaps bulging of the arc tube which may cause it to shatter.

Ballasts will overheat if lamps are operated on lower-wattage ballasts, even if they are the correct type. A ballast is designed to provide a given current. Changing lamp wattage will not change ballast current.

Lamp Operating Watts

All lamps are assigned a nominal wattage. They will operate at this wattage only if the line voltage and lamp voltage are nominal. This does not always happen, so that the actual wattage output of lamps can vary ±10 percent of nominal wattage. Variations as much as ±20 percent sometimes occur. The result is that measured illumination levels may not match calculated levels exactly.

Ballast Electrical Characteristics

All HID lamps must be operated with a ballast for the following reasons:

1. To provide the correct starting current

2. To provide the correct voltage to start the arc discharge

3. To provide the correct voltage to stabilize the arc and operate the lamp

4. To control the flow of electric current through the arc discharge

5. To compensate for the low power-factor characteristic of the arc discharge

STARTING CURRENT

Starting current is the current available to the lamp during the first 30 seconds or so of the warm-up cycle. This current must be within the tolerance range specified in the appropriate lamp standard or manufacturer's specification. If the current is too high, lamp life could be shortened. If it is too low, the lamp may not warm up to full output. A ballast which provides starting current near the maximum value will provide fast warm-up of the lamp at some sacrifice in life. Starting current near the minimum will cause slow warm-up, but tends to give longer lamp life.

The minimum allowable starting current for mercury lamps is the same as the operating current; the maximum is about twice the operating current. For metal halide and HPS lamps, the minimum current is usually the same as the operating current, and the maximum is about 50 percent greater.

STARTING VOLTAGE

Starting voltage is the open-circuit voltage required from the ballast to provide sufficient ionization in the lamp arc tube to establish current flow and an arc between the main electrodes. This causes the arc to strike and the lamp to warm up. High-pressure sodium lamps do not have an auxiliary electrode; a high-voltage pulse from a separate starting aid is therefore needed to start HPS lamps. The starting pulse needed for HPS lamps of 400 watts or less is about 2500 volts; for the 1000-watt sizes it is 3000 volts. This is sufficient to ionize the gases in the arc tube and start the arc.

The starter aid supplies this added high-frequency voltage pulse with one pulse each half-cycle of a 60-hertz wave. As soon as the arc is struck, the added pulse disappears, but it must be capable of operating over extended periods, as it must when a lamp fails or cycles on and off at the end of life.

Mercury and metal halide lamps require a higher starting voltage when started at low temperatures, because there is less vaporized mercury present in the arc tube to assist ionization. For example, a 400-watt mercury lamp requires 190 volts to start at +10°C (+50°F) and 225 volts to start reliably at −29°C (−20°F). Most ballasts for outdoor applications will start reliably at temperatures as low as −29°C, and some HPS ballasts will operate at even lower temperatures. The ballast must provide the voltage necessary to start at these lower temperatures throughout its range of line volts. Ballast literature gives information concerning minimum line voltage and minimum starting temperatures.

OPERATING VOLTAGE

Each lamp is rated by the manufacturer to operate at a certain nominal voltage. However, a new lamp can have a wide spread of actual operating voltages. This spread must be accommodated by the ballast, to minimize the variation in lamp watts.

In mercury and metal halide lamps, the original lamp voltage remains fairly constant throughout lamp life. Lamp watts, then, in these sources will vary almost directly with lamp volts and will also remain fairly constant.

On the other hand, HPS lamps contain an excess dose of mercury and sodium amalgam. Thus, as the arc tube blackens, the thermal balance slowly increases, the pressure slowly increases, and the lamp voltage rises. This voltage will gradually increase at an approximate rate of 1 to 2 volts per 1000 hours to about 140 volts near the end of lamp life for a 400-watt lamp. The ballast, then, must maintain reasonably constant wattage from the lowest new-lamp voltage of 84 to 90 volts to the end-of-life voltage of 140. Some newer, solid-state ballasts can compensate for the full range of the lamp voltage, providing almost constant lamp wattage and lumen output.

Some lamps operate at a different voltage horizontally than vertically, particularly the larger mercury lamps. This can result in a change in lamp watts in different operating positions. Manufacturer's literature should be consulted on this point.

CURRENT FLOW

Ballasts are designed to regulate the flow of current through the arc of the lamp. Current crest factor gives a picture of the shape of the current flow. *Crest factor* is the ratio of peak (maximum) to rms (root mean square) value of current or voltage waveforms. Crest factor is a function of the ballast, and not of lamps or luminaires. The crest factor of a sinusoidal waveform is 1.41, as shown in Fig. 2-6.

Figure 2-6 also shows a distorted waveform, where the ratio of peak value to rms value is higher than 1.41. This is a nonsinusoidal current waveform.

Crest factor is important in ballast design because of its effect on lumen maintenance, which is the continuing lumen output of the lamp over its life. Lumen maintenance curves from lamp manufacturers are based on operating with a sine wave of lamp current with a crest factor of about

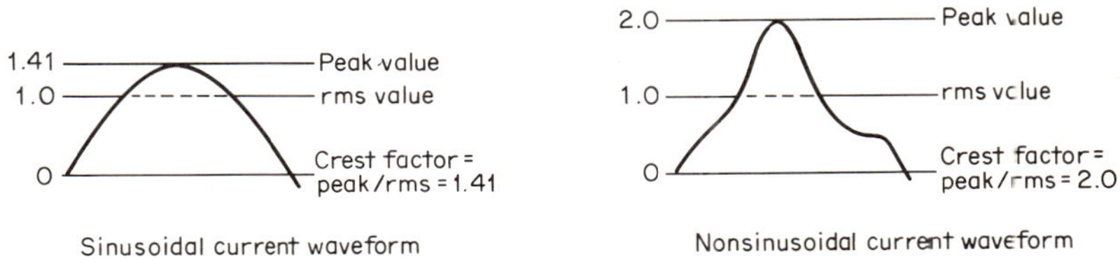

Fig. 2-6 A sinusoidal waveform has a crest factor of 1.41. A nonsinusoidal waveform has a peak value other than 1.41. The current crest factor for ballasts used with mercury lamps can be no higher than 2.0; crest factor for ballasts used with metal halide and HPS lamps should be no higher than 1.8.

1.41. A crest factor higher than that causes particles of lamp electrodes to blast off at a higher rate. This clouds the arc tube more quickly and results in lumen maintenance less than that in published data.

Manufacturers have set maximum crest factors for various types of lamps. Mercury lamps will tolerate a ballast with a crest factor of 2.0 or below; metal halide and HPS lamps must have ballasts with crest factors no higher than 1.8. These limits have been set in order to minimize variations from rated performance, which might be caused by higher current crest factors.

Voltage crest factors can be higher than 1.41 with no effect on lumen maintenance. Higher voltage crest factors are often necessary in order to provide enough peak voltage to start a lamp, as well as to keep the rms voltage needed to sustain the arc at the lowest possible level. Designing ballasts with high peak voltages can result in lower costs than making the rms higher in order to provide enough peak voltage. Alternatively, a peaking capacitor can be used to provide sufficient peak voltage while still keeping rms voltage low. Certain metal halide lamps must not be used with peaking capacitors; manufacturer's data should be consulted.

POWER FACTOR

It is advantageous to have as high a power factor as possible in the lamp-ballast combination. The power factor for mercury sources is quite high: 91 percent. However, HPS sources have a power factor of about 86 percent.

Using a correct ballast with HPS sources can compensate for the power factor of the arc discharge. Most HPS lamp-ballast combinations have power factors of 90 to 95 percent.

Ballast Operating Characteristics

There is certain standard information that manufacturers supply about their ballasts. The designer should have general understanding of the factors involved before choosing a ballast type. Table 2-2 is a typical published chart of the characteristics of ballasts designed for use with mercury or metal halide lamps.

LINE VOLTS

It is always necessary to know whether there are any restrictions on line volts when operating a certain type of ballast. Ballasts should only be connected to circuits of the voltage and frequency for which they are labeled.

Table 2-2 Ballast operating characteristics—mercury lamps

	Line Volts	Voltage Input Range	Power Factor	Starting Current	Input Voltage Dip (4 sec)	Ballast Losses	Crest Factor of Lamp Current
Reactor ballast	240 and 277 V for 100 W through 400 W, 480 V for 700 and 1000 W	±5%	50% lag (NPF)	Higher than operating current	15–20%	Low	1.41–1.5
	Same as above	±5%	90% (HPF)	Slightly higher than operating current	15–20%	Low	1.41–1.5
Lag ballast	120 V for 100 W through 400 W	±5%	50% lag	Higher than operating current	15–20%	Higher than reactor	1.41–1.5
Regulator ballast	Any voltage	±13%	95%	Lower than operating current	50–60%	Highest	1.8–2.0
Autoregulator ballast: Mercury lamps	Any voltage	±10%	90%	Lower than operating current	40–50%	Slightly under regulator	1.8–2.0
Autoregulator ballast: Metal halide lamps	Any voltage	±10%	90%	Lower than operating current	40–50%	Equivalent to mercury regulator	1.6–1.8

Otherwise, lamp performance will be different from rated values, because of changes in lamp watts, and the ballast could be damaged.

VOLTAGE INPUT RANGE

Most power distribution systems regulate their power output so that nominal line voltage varies no more than ±5 percent. Some systems, however, can vary as much as ±10 percent from nominal voltage. Some ballasts will not tolerate changes of more than ±5 percent; others will not be affected by variations of as much as ±13 percent. Ballasts must be chosen to conform to expected variations in line voltage on the particular system where they will be installed. If supply voltage dips below variations allowed by the particular ballast, the lamp may not start, lamps may start but not stabilize, or lamps may start and warm up but at reduced wattage.

POWER FACTOR

Ballasts with high power factors make more efficient use of the distribution system than ballasts with normal or low power factors. *Power factor* (PF) is the ratio of line watts to line volt-amperes,

$$\% \text{ PF} = \frac{\text{line watts}}{\text{line volts} \times \text{line current}} \times 100$$

According to ANSI standards, a ballast can be rated as "high PF" if it has a PF of 90 percent or higher. Most normal- and low-PF ballasts are around 50 percent.

A lower PF rating means more line current per ballast on a circuit, which requires larger wire size, larger switches, and a larger distribution transformer for the equivalent connected load.

A normal-PF ballast can require almost twice as much line current as a high-PF ballast. Lower initial costs for normal- or low-PF ballasts, then, are sometimes offset by increased costs of the distribution system.

STARTING CURRENT

It is necessary to know the relationship of starting current to operating current because fuses, circuit breakers, and control switches must be sized to accommodate whichever current is higher.

INPUT VOLTAGE DIP (4 SECONDS)

The input voltage dip is the dip in line voltage that a ballast will tolerate until the lamp is extinguished. The point where the lamp is extinguished is known as the *extinction voltage,* and is defined by ANSI as "the highest value of supply voltage which causes extinction of the lamp as supply voltage is reduced continuously at a rate of 2 to 3 percent per second from rated value to a point where the lamp extinguishes."

All distribution systems are subject to dips in line voltage, because of loads being switched in or out, or other factors. These dips are usually about 10 percent, but on some circuits dips can be as much as 20 to 30 percent. Most dips last for 16 to 30 cycles. Dips of a few cycles may be caused by certain heavy loads of short duration, such as resistance welders. Short dips may cause lights to blink but will seldom extinguish them.

Ballasts must be capable of riding out expected dips in line voltage without extinguishing lamps. Since most ballasts operate at 60 cycles per second, a dip of 4 seconds would be a dip of 240 cycles.

BALLAST LOSSES

The ballast losses or efficiency of a ballast can be determined from the following formula:

$$\text{Efficiency} = \frac{\text{lamp watts}}{\text{lamp watts} + \text{ballast losses}} = \frac{\text{output watts}}{\text{input watts}}$$

A ballast that is 90 percent efficient delivers 90 percent of the power to the lamp, and only 10 percent to heat the ballast. Ballast losses are measured on a bench in order to obtain repeatable data. When a ballast is enclosed in a housing the stray losses increase approximately 3 percent. The watt losses of a ballast add to the total power consumed, so the designer must consider losses when choosing a ballast.

Losses are difficult to measure accurately, so ballasts are often rated on a comparative basis, as in Table 2-2. Usually efficiency is greater with a higher lamp-wattage rating.

CREST FACTOR OF LAMP CURRENT

Crest factor was explained earlier. A sine wave has a crest factor of 1.41. Mercury lamps must have a ballast with a crest factor of 2.0 or below. Metal halide and HPS ballasts must have crest factors no higher than 1.8.

Ballasts for Use with Mercury Lamps

There are several basic types of mercury ballasts, as shown in Fig. 2-7. Information is given in this chapter for the ballasts of one manufacturer, but other manufacturers' products are similar and are based on the same operating and design principles.

RESISTOR BALLAST

A plain resistor ballast should only be used on dc systems, with mercury lamps, since it is sensitive to dips in line voltage. The resistor ballast limits line current, but its efficiency is only about 50 percent because high power levels are dissipated in the resistor. Normally only mercury lamps operate stably on a dc source, since the plasma components in metal halide lamps tend to separate and migrate under dc operation. This is the simplest type of ballast.

REACTOR BALLAST

The reactor ballast is the simplest magnetic ballast for use on ac circuits. It is a choke coil wound on an iron core and placed in series with a lamp, operated directly from a line of suitable voltage.

Reactor ballasts can be used with or without an across-the-line capacitor. The purpose of the capacitor is to increase the PF, which is only 50 percent for a plain reactor, but goes to more than 90 percent with the addition of a PF-correction capacitor.

Advantages The reactor ballast is small, lightweight, inexpensive, and efficient. It provides good regulation of lamp watts with variations in lamp volts; a 3 percent change in lamp watts occurs with an 11 percent change in lamp volts. The ballast provides high starting current for fast warm-up.

Disadvantages A ±5 percent change in line volts will result in a variation of ±10 percent in lamp watts. The ballast can be used only on spec-

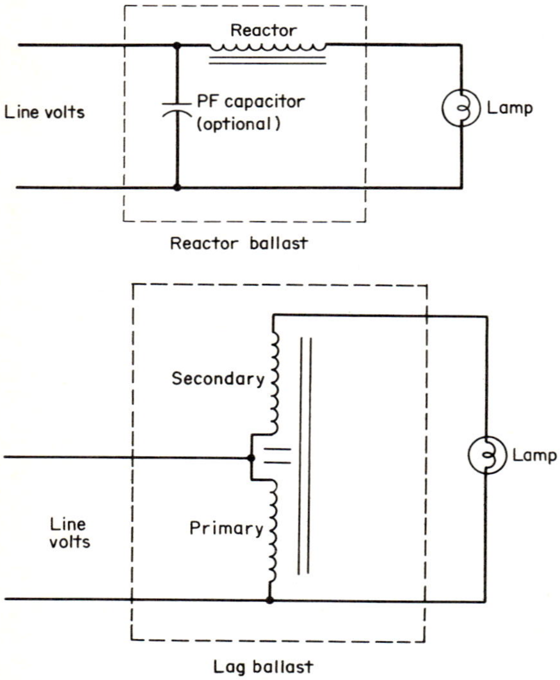

Fig. 2-7 Ballast used with mercury lamps. The electric circuit of the metal halide peak-lead ballast is the same as that for the mercury autoregulator. However, the peaklead ballast has core slots which produce a highly peaked waveshape of open-circuit voltage.

ified line voltages: usually 240 or 277 volts for 100-watt through 400-watt lamps, and 480 volts for 700- and 1000-watt lamps. Use of the reactor ballast is limited to circuits where ±5 percent of line voltage is maintained, for acceptable regulation and cold-weather starting. As mentioned, PF is low unless a capacitor is added to the ballast.

LAG BALLAST

The lag ballast combines a reactor and an autotransformer. The ballast's electric circuit is shown in Fig. 2-7.

This ballast is made chiefly for operation on 120-volt circuits, with no capacitor added to correct the 50 percent PF. It is seldom used in a high-PF capacity, because the added capacitor makes it as expensive as ballasts which have better performance.

Advantages The lag ballast has the same good features as the reactor ballast: good regulation of lamp watts with variations in lamp volts; low current crest factor; and high starting current to the lamp. In addition, it

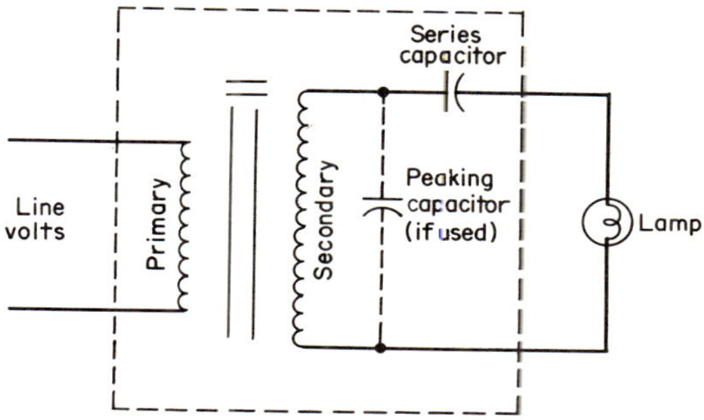

Regulator ballast

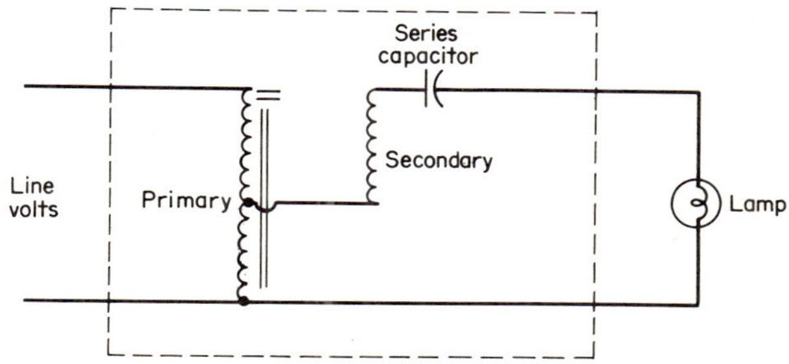

Autoregulator ballast

Fig. 2-7 *Continued*

can be used when line voltage is lower or higher than the required starting voltage of the lamp.

Disadvantages The lag ballast is larger than a reactor ballast, and thus more expensive. It is also less efficient.

REGULATOR BALLAST

The mercury and metal halide regulator ballast consists of a capacitor in series with a lamp and either in series or parallel with two separate windings electrically isolated from each other (see Fig. 2-7). In some designs a peaking capacitor can be added to mercury regulators to increase open-circuit voltage for low-temperature starting. The secondary core operates in magnetic saturation, so that secondary current remains relatively constant over a wide range of primary voltage variations. This results in regulation and stabilization of lamp wattage.

Advantages The regulation provided by this ballast is such that line-voltage changes of ±13 percent result in only ±2 or ±3 percent changes in lamp watts. As seen in Table 2-2, PF is high: 95 percent. The ballast can be installed on any circuit, with little concern for line-voltage variations or voltage dips. It can be used at any line voltage, and dual-voltage or multiple-voltage ratings are available.

Disadvantages The regulator ballast is more expensive than the reactor ballast. The current crest factor is also somewhat higher.

AUTOREGULATOR BALLAST

The autoregulator ballast combines a regulator circuit with an autotransformer (see Fig. 2-7). Part of the primary winding is coupled with the secondary winding, thus reducing ballast size. Only the secondary winding contributes to the regulation of lamp watts with variations in line volts, and so the degree of regulation depends on the amount of primary winding which is common to the secondary.

Advantages The autoregulator has the same advantages as the regulator ballast: high PF, tolerance to dips in line voltage, and low line starting current. In addition, it is smaller, lighter, and less expensive than the regulator. Losses are somewhat lower. Regulation is high, though slightly lower than that of the regulator. Line voltage changes of 10 percent can result in changes in lamp watts of about 5 percent.

Disadvantages The autoregulator does not provide isolation between primary and secondary windings, so regulation capability is reduced somewhat from that of the regulator ballast.

TWO-LAMP BALLASTS

Two-lamp ballasts can be used in either series operation or parallel operation, with one or more of the basic circuits, either singly or in combination. Series designs offer the least expensive system, together with reduced size and weight. However, both lamps go out when one lamp fails. Two-lamp parallel designs have two independent ballast circuits. If one lamp fails, it does not affect the performance of the second lamp.

Ballast life is not affected if both lamps are removed, but removal of only one lamp can affect ballast life and may void the product warranty.

Two-lamp parallel designs can be of regulator or lead-lag design. Lead-lag two-lamp ballasts have high PFs. Also, stroboscopic effect is not apparent when these ballasts are used with mercury lamps. A disadvantage of these ballasts is that they will not tolerate line-voltage variations of more than ±5 percent.

Ballasts for Metal Halide Lamps

A special ballast has been developed for use with metal halide lamps. Even though metal halide lamps are essentially like mercury lamps, there are two significant differences:

1. A higher ballast open-circuit voltage is needed to start the arc at a specified temperature.
2. A higher reignition voltage is needed.

These two differences occur because of the nature of the ionization phenomena of the metallic additives that are present in metal halide lamps. During the warm-up period of the lamp, low conduction of the arc occurs. This means that the reignition voltage needed to establish or reestablish an arc after each half cycle is greater than the voltage a standard mercury ballast can normally supply.

There is one line of metal halide lamps which has been designed specifically for use with mercury ballasts. Other metal halide lamps should not be operated on mercury ballasts, however, because the arc will continually be extinguished, and have to cool and restart. This may not be noticeable in the beginning, but will be very pronounced after a few thousand hours of operation.

The ballast for use with metal halide lamps is called *peak-lead* or *lead-peaked*. The electric circuit is the same as a mercury autoregulator, as shown in Fig. 2-7. The difference between the two is that the peak-lead ballast contains one or more core slots on the secondary winding. These core slots produce a peaked waveshape of open-circuit voltage. The peak voltage is sufficient to start the lamp, and also allows the rms value of the open-circuit voltage to be less than would be necessary if the wave were not peaked and had to be extended to allow enough voltage to start the lamp.

This design is different from the design of a mercury ballast with a peaking capacitor. Metal halide lamps should not be operated on mercury ballasts with peaking capacitors, because they will cycle on and off. However, all *lead-peak* ballasts will operate mercury lamps having characteristics similar to metal halide lamps of the same wattage.

The crest factor of the metal halide ballast varies between 1.6 and 1.8. Regulation is between that of a regulator ballast and that of a lag or reactor ballast. The metal halide ballast also provides low line starting current, tolerance of line voltage dips of 40 to 50 percent, and high PF.

Ballasts for High-Pressure Sodium Lamps

An HPS lamp requires a ballast much different from the ballasts used for other HID sources. The differences are neccessary for two reasons:

1. The voltage required to start the lamp is much greater.
2. Lamp voltage increases over the life of the lamp.

Starting voltage in an HPS ballast is provided by an auxiliary starting circuit. This supplies a high-voltage pulse of 2500 volts for lower-wattage sources and 3000 volts for the 1000-watt lamp.

Lamp voltage increases over the life of the lamp because of the slow increase in the stabilized temperature of the arc tube. As the lamp ages and becomes hotter, more and more amalgam is vaporized, resulting in increased lamp voltage. The HPS ballast must be designed to hold the lamp wattage within certain limits, as defined by the lamp manufacturer.

The Amercian National Standards Institute sets parameters within which an HPS lamp must operate. A 400-watt source operating at 100 volts is required to operate within a range of 84 to 140 volts. The lamp watts can vary from 280 to 475. This means that lamp volts can vary as much as +15 percent and −16 percent. Lamp watts can vary as much as +19 percent or −30 percent from nominal. This can result in wide changes in wattage and light output if ballasts with tight controls are not used.

High-pressure sodium ballasts have an operating curve which varies with ballast design. When a ballast is operated with a 400-watt source, its curve is not usually shown below 80 volts or 300 watts, since this is not part of the operating curve but is used only during warm-up. The *x* or dot at the end of a ballast curve shows the point where voltage available from the ballast is not enough to keep the lamp operating. At that point the lamp will begin to cycle off and on.

Figure 2-8 compares the performance of two different ballast designs with a 400-watt HPS lamp. The normal manufacturing range of this source is 90 to 115 volts, which is labeled *lamp design center* on the drawing.

Any ballast used with this lamp must be able to operate the lamp from 84 to 140 volts, since the lamp could be dropping 140 volts by the end of life. The lamp wattage range could be from 280 to 475 watts and the ballast should be able to handle this variation as well. These two parameter combinations form a trapezoid as illustrated in Fig. 2-8.

The design 1 ballast in the drawing is a fairly typical HPS ballast design. The light output of the system starts below nominal value, so that it will increase to nominal value as the lamp ages. The increased lamp wattage

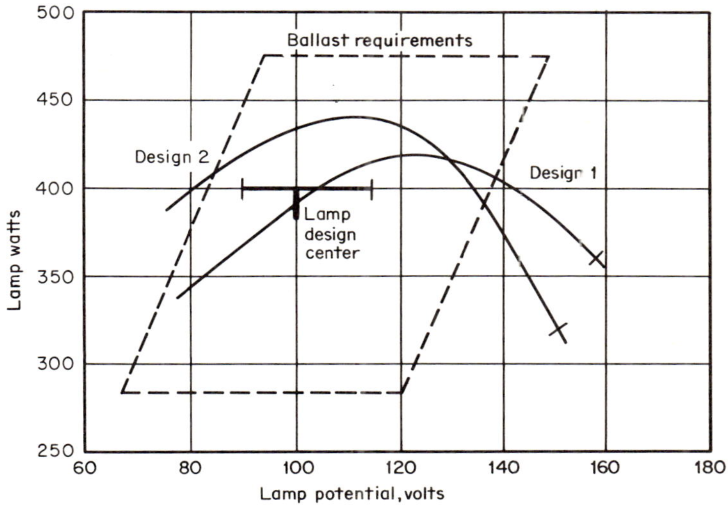

Fig. 2-8 Ballast performance can be plotted on the trapezoid of ballast requirements for HPS lamps. In this case the lamp is 400 watts, with a normal range of 90 to 115 volts. The ballast Design 1 comes closest to a steady output of 400 watts and is the better choice.

common with HPS lamps compensates for lamp depreciation, so that light output increases. With this ballast, the average light output of the lamp will remain fairly constant over life.

The design 2 ballast will not provide lamp performance as good as that provided by design 1. Light output from the system will go down as the lamp ages, particularly toward the end of life.

STANDARD HPS BALLASTS

Characteristics of the four types of HPS ballasts are shown in Table 2-3. It should be noted that in each case tolerance for lamp dips is rated at nominal lamp volts; as the lamp ages, the dips that a ballast will tolerate constitute a smaller percentage of actual lamp volts.

Reactor Ballast A drawing of the electric circuit for a reactor ballast is shown in Fig. 2-9. This standard ballast can be used for HPS lamps if a starting aid is added and open-circuit voltage is correct for the lamp.

Reactor ballasts of 70, 100, and 150 watts (55 lamp volts) can be used directly on a 120-volt system. With an autotransformer, the ballasts can be used on 208-, 240-, 277-, and 480-volt lines. Reactors of 250 and 400 watts can be used directly on a 208-volt system, but with an autotransformer they can also be installed on 120-, 240-, 277-, and 480-volt lines. The 1000-watt reactor ballast can be used only on a line of 480 volts.

Figure 2-9 shows how one reactor-ballast design would work with an

Table 2-3 Ballast operating characteristics—high-pressure sodium

	Line Volts	Voltage Input Range	Power Factor	Starting Current	Input Voltage Dip (4 sec)	Ballast Losses	Crest Factor of Lamp Current
Reactor ballast	Any voltage (furnished complete with autotransformer)	±5%	90%	Higher than operating current	15–20%	Equivalent to mercury regulator	1.4–1.5
Magnetic regulator ballast	Any voltage	±10%	95%	Lower than operating current	40–50%	Highest	1.7
Autoregulator ballast	Any voltage	±10%	95%	Lower than operating current	40–50%	Equivalent to HPS reactor	1.6–1.8
Electronic regulator ballast	Any voltage	±10%	85–90%	Higher than operating current	20–40%	Lower than magnetic or autoregulator	1.6

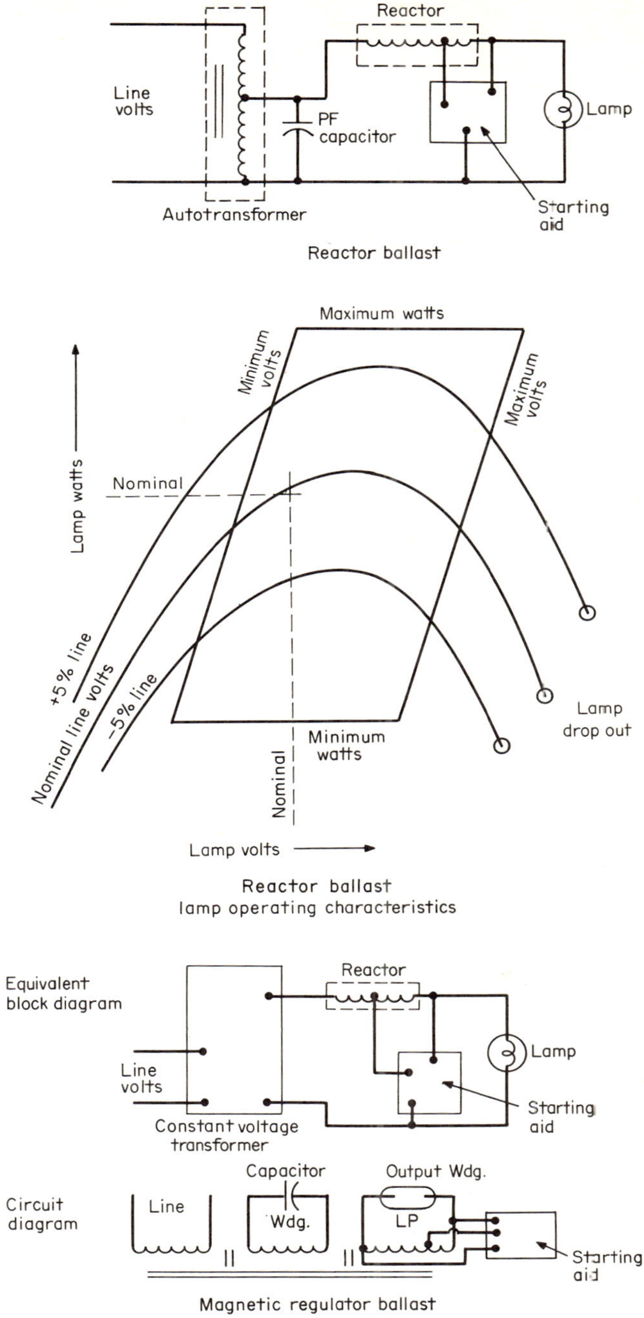

Fig. 2-9 Electric circuits for three types of HPS ballasts, and the operating characteristics of the lamp when it is used with each type. In each case, the lamp will operate best at nominal line voltage.

HPS lamp. At nominal voltage, the lamp operates at nominal wattage; at +5 percent voltage, it operates above nominal wattage; and at −5 percent voltage, it operates below nominal wattage.

Magnetic Regulator Ballast A standard mercury regulator ballast cannot be used for HPS lamps, because lamp watts increase almost in direct proportion to the change in lamp volts with such a ballast. This would substantially shorten HPS lamp life.

Instead, a magnetic regulator is used, in order to compensate for

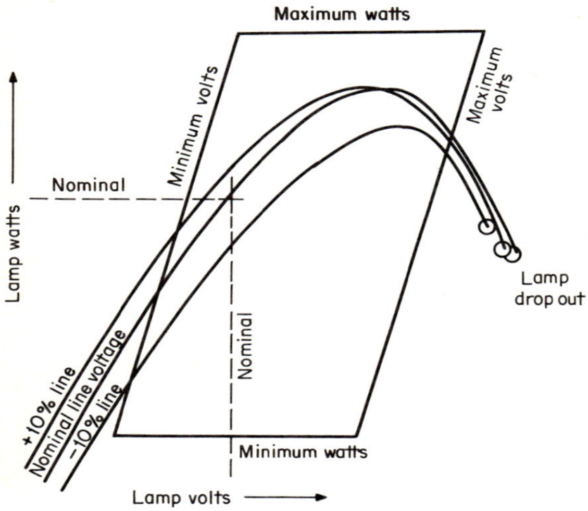

Magnetic regulator ballast lamp operating characteristics

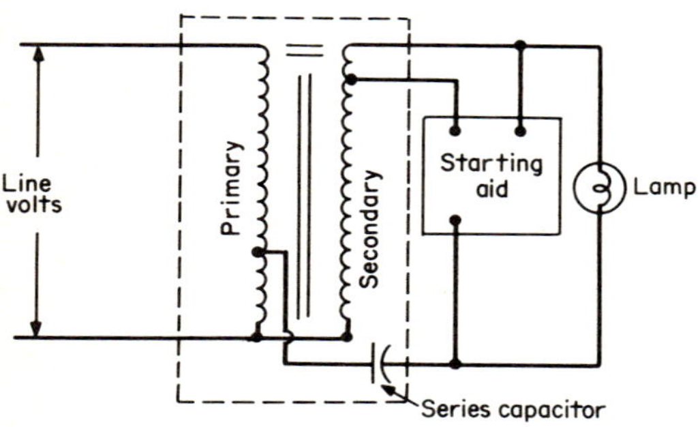

Autoregulator ballast

Fig. 2-9 *Continued*

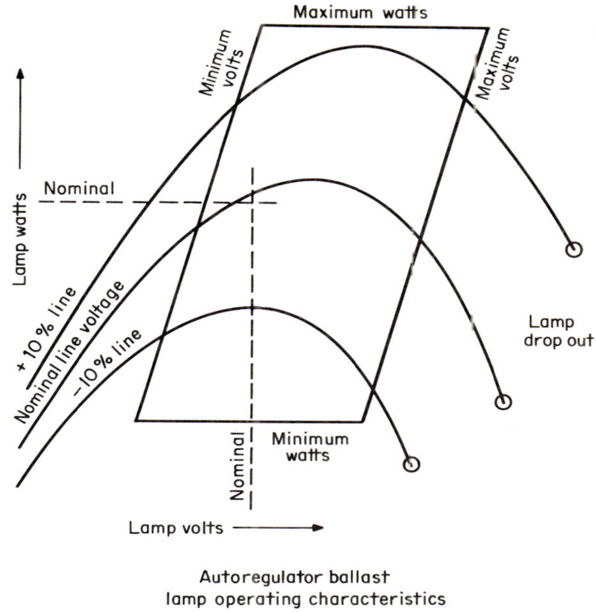

Fig. 2-9 *Continued*

changes in lamp voltage and to correct variations in line voltage. The drawing for the electric circuit of this ballast is shown in Fig. 2-9. The design is equivalent to a voltage-regulating transformer supplying the lamp through a reactor, with all elements in a single core and coil.

Lamp watts are controlled more closely with this ballast than with the reactor, as illustrated in Fig. 2-9. Variations of ±10 percent from nominal line voltage do not result in as much variation in wattage as variations of ±5 percent with a reactor ballast.

Autoregulator Ballast The electric circuit and the operating characteristics of an autoregulator HPS ballast are shown in Fig. 2-9. The ballast is like a mercury autoregulator, except that it has a much different reactance to capacitive reactance ratio.

ELECTRONIC REGULATOR BALLAST

Electronic regulator ballasts are available for HPS sources of 70 through 400 watts. This kind of ballast combines standard ballast methods with solid-state components. It incorporates a feedback loop which measures lamp volts and lamp current. The control circuit monitors this data and maintains the output of the ballast at predetermined levels, depending on the input data. As a result, the ballast operates the lamp at ±1 percent wattage variation. The ballast circuit and operating characteristics are shown in Fig. 2-10.

Allowable line voltage variation for the electronic regulator ballast is ±10 percent (see Table 2-3). Power factor is 85 to 90 percent. The ballast will tolerate line voltage dips of 20 to 40 percent.

Advantages The electronic regulator ballast provides better lamp performance than other HPS ballasts, and has other advantages as well.

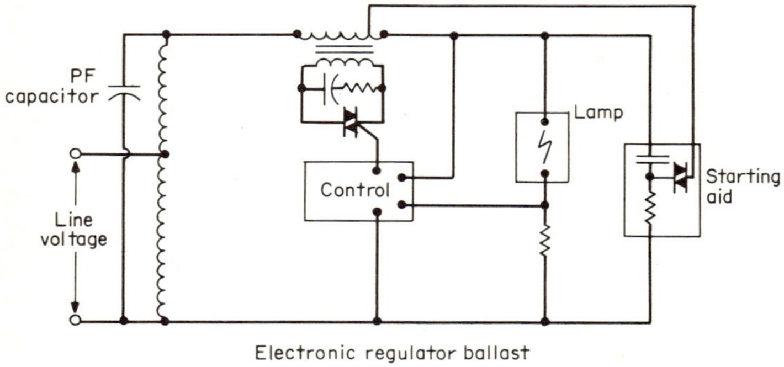

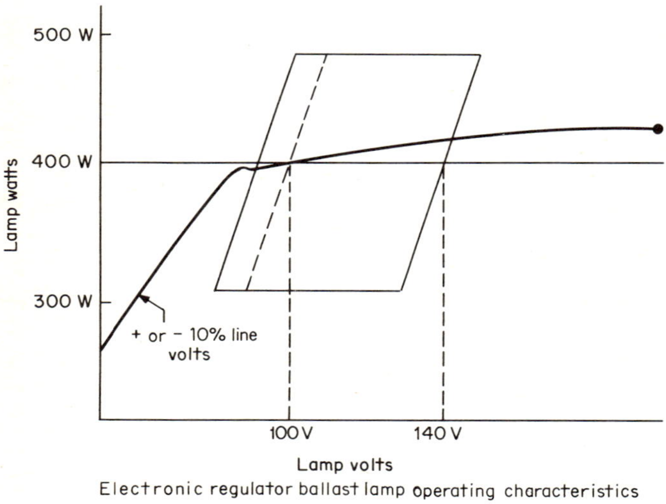

Fig. 2-10 Design of an electronic regulator ballast and graph of lamp operating characteristics. The ballast incorporates a feedback loop, which measures lamp volts and lamp current. The control circuit monitors the data and controls the output of the ballast, so that lamp wattage will vary only ±1 percent. This means that light output at the end of lamp life will be almost the same as at the beginning of life.

1. **Control.** The electronic regulator ballast operates a lamp within ±1 percent at any point in time, within the following variations: input-line voltage variations of ±10 percent; initial lamp volts of 85 to 115 volts; capacitor variations of ±3 percent; and variations of ±5 percent for all manufacturing tolerances.

2. **Lamp Lumen Depreciation and Levels of Illumination.** One electronic regulator ballast compensates for lamp lumen depreciation (LLD). As a result, the LLD factor for an HPS lamp using this ballast is 1.0; that is, the lumen output at the end of life is nearly equal to the initial lumen output. This ensures the accuracy of lighting design. At the end of life, HPS lamps using electronic regulator ballasts will be producing almost the same level of illumination as they did at the beginning of life.

3. **Number of Luminaires.** Because lamp lumens are nearly constant throughout lamp life, fewer luminaires need to be used in an installation — often as many as 10 percent fewer. Since LLD is 1.0 when the electronic regulator ballast is used, rather than the usual 0.9 or lower, the number of luminaires in a given installation can be less and still produce the required level of illumination.

4. **Operating Costs.** Because of superior control of watt output, there is lower wattage consumption per luminaire with an electronic regulator ballast. This means lower energy and operating costs. In addition, operating costs are reduced because fewer luminaires are used in the installation.

Disadvantages Electronic regulator ballasts cost more than standard HPS ballasts, but this may change as their use becomes more widespread.

NEXT GENERATION OF BALLASTS

Development has begun on another type of ballast, though none has been made commercially as yet. This ballast will be able to control the light output of an entire lighting system. At the time of this writing, it is necessary to have a separate system for each ballast if dimming is desired. This increases system costs.

The next generation of ballasts will have a control circuit directed by remote control in conjunction with stepped output control in the ballast. The ballast will be able to cut lamp watts to three-fourths, one-half, or one quarter of rated wattage. This will ensure greater flexibility in the lighting system, allowing lamp watts to be changed in fixed amounts for different uses and changing activities.

Kinds of Housings

Ballasts are generally available in a simple core-and-coil arrangement, in a linear housing, or encased.

CORE-AND-COIL BALLASTS

Core-and-coil ballasts are designed for in-head application, and may be installed in any position. They are small and lightweight.

The insulation system is designed for a maximum 180°C hot-spot temperature, or 165°C maximum observable temperature, as determined by the change-of-resistance method.

The temperature rise of the ballast coils, plus the maximum ambient temperature, must not be greater than 165°C. For instance, if the ambient temperature were 40°C, then the temperature rise of the coil could not be greater than 125°C. In very hot industrial applications, such as heat-treating areas, boiler rooms, and foundries, the ambient temperature is often as high as 65°C. In these cases, ballast-coil rise cannot be more than 100°C.

LINEAR BALLASTS

Linear mercury ballasts can be encapsulated in standard fluorescent ballast cases. These should be installed so that ballast-case temperature does not exceed 90°C at the hottest spot once the ballast has reached its maximum operating temperature. These ballasts may be remotely mounted indoors.

ENCASED BALLASTS FOR INDOOR USE

Encased ballasts are intended for indoor use and should not be used outdoors unless they are protected from weather conditions. Louvered cases should not be used in atmospheres containing conductive dust.

Temperatures on the cases should not exceed 90°C. If ambient temperatures are extremely high, specially rated ballasts should be used. Manufacturers should be consulted.

Remote Mounting of Ballasts

For HPS lamps, the maximum distance from a core-and-coil ballast to the lamp is 1.78 meters (5 feet 10 inches) for 70-, 100-, and 150-watt lamps. For 250-, 400-, and 1000-watt lamps, the maximum distance is 1.5 meters

(5 feet). Special starting aids need to be used if 250-, 400-, or 1000-watt lamps are to be mounted more than 1.5 meters (5 feet) from core-and-coil ballasts.

Encased ballasts can be mounted 10.7 meters (35 feet) from HPS lamps. This is the maximum recommended distance.

Grounding and Wiring

All ballasts must be grounded to avoid shock hazard, personal injury, and damage to luminaires. Electric grounds should comply with the *National Electrical Code* and the appropriate local electrical code or ANSI standards.

Line Fusing

Most electrical equipment is fused for fire protection, or protection of equipment and wiring. Other reasons include preventing release of vapors or objectionable liquids, or isolating faulty equipment before it trips a branch circuit breaker.

Most of these reasons are not applicable to HID ballasts. The principle reason for fusing an HID ballast is to remove a faulty luminaire from a circuit before it results in a branch-circuit outage. This is helpful in high bays or other relatively inaccessible areas.

Fuses decrease the reliability of ballast systems and add to costs. However, if line fusing is done, fuses must conform to the required voltage rating and must be sized to the ballast manufacturer's recommendation.

Radio Interference

There may be some radio interference as the arc is being established in a lamp. There will be no objectionable radio interference during normal operation.

Noise

All HID ballasts generate some noise. This is usually not a problem in industrial areas. However, if such noise is objectionable, then sound-absorbing techniques should be considered.

There is some noise from HPS ballasts during starting, or if the starting aid operates continously because the lamp has failed and is attempting to start. (See also "Lighting System Noise Criterion" in Chap. 3.)

Auxiliary Lighting

Most ballast units are available with a 120-volt lead, so that an incandescent auxiliary lamp can be used with the luminaire. This is often necessary in industrial areas, because of the warm-up time required for HID sources. Sometimes power is extinguished briefly; when the power returns, incandescent lighting can be used while the HID sources restrike.

A special switch must be used so that the HID lamp and the auxiliary lamp don't operate at full wattage at the same time. If they do, the ballast life will be shortened.

Dimming Systems

Dimming systems are now available for all major types of HID lamps. Most systems use the principle of sine-wave phase control. In order for the lamp to keep operating the same peak voltage must be supplied to the lamp. What the dimming system does is to supply this voltage for a shorter or longer period of time. This results in a lower or higher wattage being supplied to the lamp, so that it operates in a dimmed mode or at increased light output.

Dimming equipment can be controlled manually, by timers or computer interface, or automatically. A dimmer with automatic controls monitors levels of illumination and adjusts them automatically to a preset level. Such a system usually uses a photocell sensor to determine the average room illumination level. If the average is above or below the preset level, the system will automatically adjust light output.

Automatic dimming systems operate more effectively if group relamping is used, rather than spot relamping. The photocells actually monitor only a few typical areas, and if a lamp burns out in an area that is not monitored, the photocells will not respond.

ADVANTAGES

Dimming systems allow light levels to be adjusted to a certain preset level. Thus, as lamps deteriorate and dirt collects on surfaces, the dimming system will respond and increase light output. Conversely, the system can be dimmed when the normal level of illumination is not needed. This can result in energy savings.

DISADVANTAGES

Initial costs for a dimming system are higher than those for a standard system, because each lamp needs to have a separate dimming ballast, which is more expensive than a standard ballast.

If an HID lamp is dimmed too far, the arc may be lost. Color stability is not always maintained with a dimming system. As input voltage drops, the temperature of the ionizable gas in the lamp also drops. This can cause a color shift.

Bibliography

Ballasts for High Intensity Discharge Lamps, Ratings and Data, GEA-8923C, General Electric Company, Lighting Systems Department, Hendersonville, N.C., 1977.

"Basics of Solid-State Lighting Controls—Part 4," *Lighting Design and Application,* October 1977, pp. 30–31.

Beardsley, Charles W.: "Let There Be Light, but Just Enough," *IEEE Spectrum,* December 1975, pp. 28–34.

Clarkson, Clarence W.: "Through a Glass Lightly," *Lighting Design and Application,* December 1973, pp. 48–51.

Collins, B. R., and S. A. Mule: "HPS Lamp Design—a Diversification in Design Wattage and Optical Control," *Lighting Design and Application,* March 1977, pp. 26–31.

Collins, B. R., W. E. Smyser, and J. T. Suter: "Low-wattage, Low-voltage HPS Lamp Design and Application," *Lighting Design and Application,* April 1975, pp. 30–36.

Frier, John P., and A. J. Henderson: "Stroboscopic Effect of High Intensity Discharge Lamps," *Journal of the Illuminating Engineering Society,* October 1973, pp. 83–86.

"Have HID Dimming Systems Arrived?" *Factory Management,* April 1977, pp. 17–19.

High Intensity Discharge Lamps, TP-109R, General Electric Company, Lamp Business Division, Cleveland, Ohio, 1975.

Hughes, W. B.: "Why Fuse HID Ballasts?" *Plant Engineering,* March 21, 1968, pp. 88–89.

Lenz, J. E.: "Interrelated Lamp and Ballast Characteristics," presented at the HID Ballast Orientation Course, Session 2, Hendersonville, N.C., March 1974.

"Light Sources," *IES Lighting Handbook,* 5th ed., pp. 8-33–8-40, Illuminating Engineering Society of North America, New York, 1972.

Longenderfer, John: "Lighting Control—New Business for Contractors," *Electrical Contractor,* February 1976.

Loytty, Eric: "A New Arc Tube for HPS Lamps," *Lighting Design and Application,* February 1976, pp. 14–17.

McGowan, Terry K.: "High-pressure Sodium—Ten Years Later," *Lighting Design and Application,* January 1976, pp. 45–51.

Palko, Ed: "Primer on High Intensity Discharge Lamps," *Plant Engineering,* May 4, 1972, pp. 73–76.

Peel, Bill: "Automating HID Dimming Systems," *Lighting Design and Application,* June 1977, pp. 16–18.

Spira, Joel: "The Case for Dimmers," *Lighting Design and Application,* May 1974, p. 63.

Subcommittee on Photometry of Light Sources of the Testing Procedures Committee of the IES: "IES Approved Method for Life Testing of High-intensity Discharge Lamps," *Journal of the Illuminating Engineering Society,* October 1974, pp. 66–69.

3

Luminaires and Their Functions

Luminaires are designed to perform a number of functions. They provide electric connection to the lamp. They adapt the lamp and ballast to the environment, even if it is wet, corrosive, or hazardous. And most importantly, they provide a physical appearance and light distribution pattern suitable for the application where they will be used. Because of the variety of needs for lighting equipment there are thousands of different types on the market.

This chapter will discuss reflector design, performance, and materials, as well as general information about the other parts of a luminaire. Some specific points which the designer must consider when choosing a luminaire also are covered.

General Classifications

Luminaires can be loosely divided into indoor and outdoor types, and into those with fixed aiming and those which need to be aimed. Luminaires with fixed aiming include those generally used indoors, as well as some roadway, area, and high-mast designs.

Detailed information on photometric performance and design procedures for indoor luminaires is given in Chaps. 4, 5, and 7. Floodlights need to be aimed and require more complex lighting design techniques. Information on this area of lighting design is given in Chaps. 9 and 10. Information on photometric performance and design procedures for roadway, area, and high-mast luminaires is given in Chaps. 11 and 12.

Indoor luminaires can be used outdoors if they are weatherproofed. Outdoor luminaires can be used indoors with the advice of the manufacturer. Outdoor luminaires are normally rated for a maximum ambient temperature of 25°C (77°F). It may be necessary to use a separate indoor ballast instead of the built-in ballast when outdoor luminaires are used indoors, since over-temperature operation will severely affect ballast life.

Light Control

Luminaires are designed to control light distribution, and so the methods used to control light should be understood. These are based on the laws of reflection and refraction.

SPECULAR REFLECTION

The law of regular reflection, or specular reflection, applies when light strikes a flat, highly polished surface. The law states that the angle of reflection is equal to the angle of incidence. This is illustrated in Fig. 3-1. The angle of the incident ray and the angle of the reflected ray are both measured from a line called the *normal,* which is perpendicular to the surface.

When the reflecting surface is curved, the incident ray and reflected ray make equal angles with the normal to the tangent line drawn through the point of incidence. This is also illustrated in Fig. 3-1.

SPREAD REFLECTION

If a flat, polished surface is made slightly rough in some manner, such as etching, pebbling, faceting, etc., its reflection characteristics are changed. The surface then consists of small specular surfaces in irregular planes. The effect is to spread the reflected rays of light.

The amount of spreading depends on the material used, the contour of the surface, and the finishing method. In each case, though, the rays tend to be reflected in directions close to the line where specular reflection would occur.

DIFFUSE REFLECTION

Nonmetallic surfaces composed of small crystals or paint pigment particles reflect light in a diffuse manner. Rays are reflected in all directions regardless of the direction of the incident light.

REFRACTION

Transparent materials have different optical densities. When light enters a substance of different density, it will bend, unless the angle of incidence

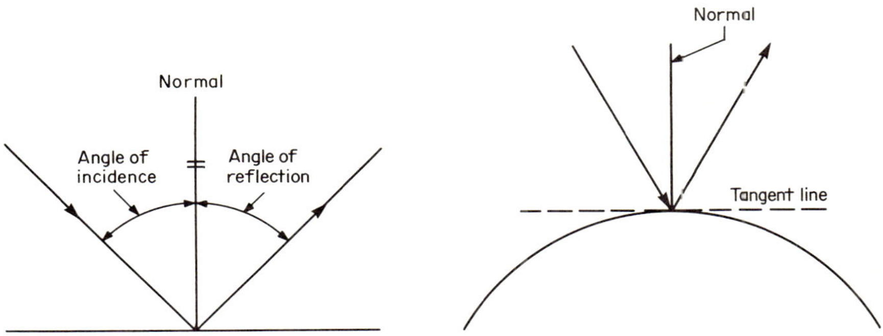

Fig. 3-1 The law of specular reflection states that the angle of incidence of the light ray is equal to the angle of reflection. When a light ray strikes a curved surface, the angle of incidence equals the angle of reflection as measured from a normal drawn perpendicular to the tangent line drawn through the point of incidence.

is the angle normal to the surface. The degree of bending depends on the relative densities of the two substances, the wavelength of the light, and the angle of incidence. Light is bent toward the normal to the surface when it enters a more dense medium and away from the normal when it enters a less dense medium.

This bending occurs at the point where the light ray enters or leaves the medium of different density. The amount of bending can be calculated by using Snell's law:

$$n_1 \sin i = n_2 \sin r$$

where n_1 = index of refraction of the first medium
i = angle the incident light ray makes with the normal to the surface
n_2 = index of refraction of the second medium
r = angle the refracted light ray forms with the normal to the surface

This is illustrated in Fig. 3-2.

Assume that n_1 is air. The index of refraction for air is 1, so for a ray traveling from air to a denser medium the formula can be written

$$\sin i = n_2 \sin r$$

Assume that the ray moves from air to glass, which has an index of refraction of 1.5. Thus n_2 = 1.5. If the incident angle i is 45 degrees, sin i is 0.707. To solve for r:

$$0.707 = 1.5 \sin r$$
$$\sin r = 0.47$$
$$r = 28.12 \text{ degrees}$$

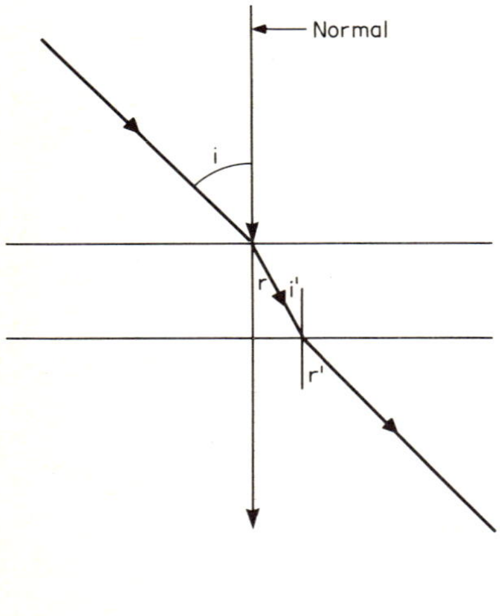

Fig. 3-2 Light striking a surface of a medium of different density will be bent or refracted. The angle of refraction follows Snell's law: $n_1 \sin i = n_2 \sin r$, where n_1 is the index of refraction of the first medium and n_2 is the index of refraction of the second medium.

If the index of refraction n_1 is greater than n_2, total internal reflection can occur. This is the principle behind the reflecting prism. This internal reflection can occur whenever sin $r \geq 1$. Sin r will equal 1 when sin $i = n_2/n_1$.

For glass to air, sin $i = 1/1.5$, and so $i = 41.8$ degrees. For incident angles larger than 41.8 degrees, light rays will be totally reflected internally.

This is the principle used for glass and plastic reflectors and for brightness control of indoor and outdoor luminaires.

Parts of a Luminaire

Reflectors are the most common means of controlling light. They can be open or a glass cover, refractor, or diffuser can be used to enclose or seal the luminaire. Louvers and visors can also be added to control stray light.

Other parts of a luminaire include a lamp socket, a ballast and terminal board, perhaps a photoelectric control, and various filtering and gasketing devices which will vary from one manufacturer to another. Figure 3-3 shows the parts of a floodlight and of a typical industrial luminaire.

All luminaires have built-in devices that allow the luminaire to be mounted in its normal orientation. There are numerous mounting adapters that can be used to fit most requirements. Luminaires can be prewired with a cord and plug to simplify installation and maintenance. Floodlights are equipped with aiming sights and degree markers to aid in their proper positioning on the job.

Every high-intensity discharge (HID) lamp requires a matching ballast. Some light sources are designed to be self-ballasting, but most are used with a ballast within the luminaire housing or remotely mounted. Information on lamps and ballasts is included in Chap. 2.

Reflector Design

The purpose of the reflector is to redirect the luminous intensity of the light source. The light distribution of the bare lamp is seldom suitable unless a reflector or some other control method has been built in to the bulb. In most cases the lamp and an external reflector are combined to produce the desired results. Since the lamp has a distribution of its own, the reflector must be designed with this in mind.

A reflector therefore becomes a secondary light source. This is obvious in a highly polished, flat reflector which can reproduce an image of the lamp itself.

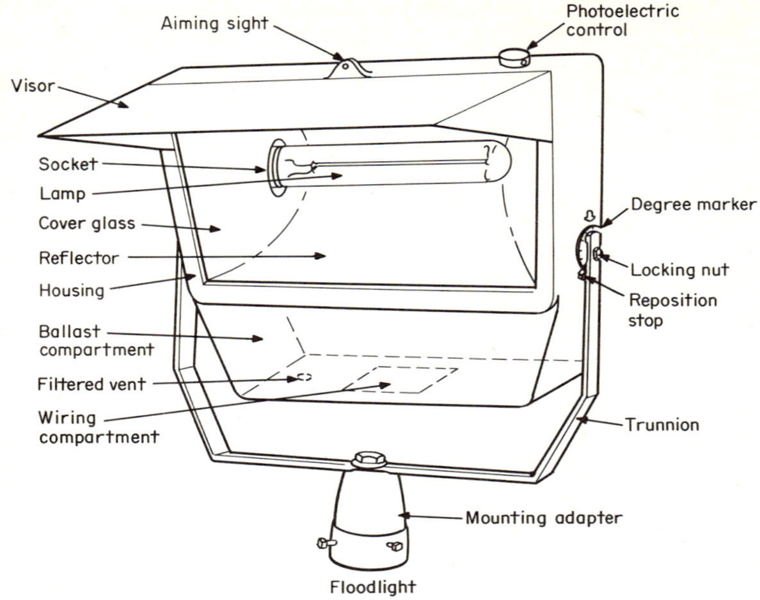

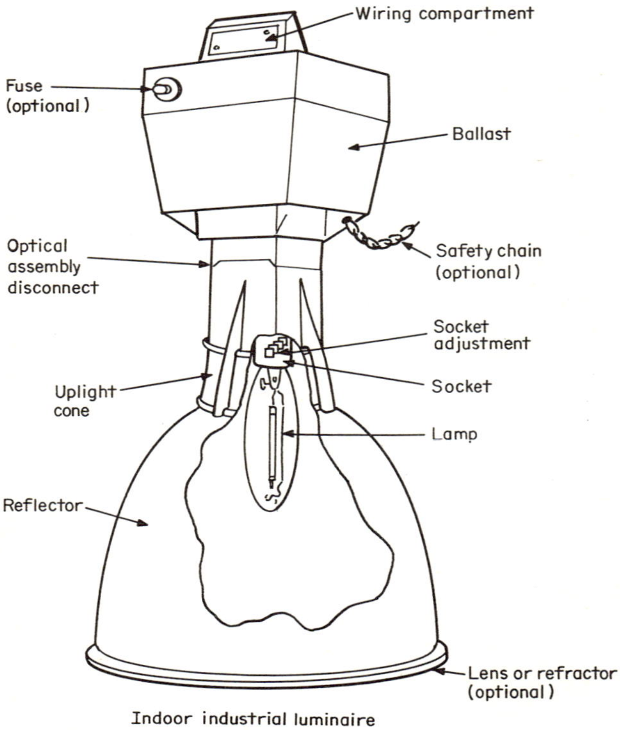

Fig. 3-3 The parts of a floodlight and of an indoor industrial luminaire.

Industrial reflectors are usually designed so the lamp can be located in more than one position to provide a range of beam spreads, although the range of beam spreads possible from one reflector is limited. If a reflector is designed for very narrow distribution and the lamp is then moved to provide very wide distribution, there will probably be a hole in the center of the beam which will produce a doughnut-shaped pattern of light.

For this reason, there are usually two or more reflector sizes, each with a range of beam spreads to fill the lighting design objectives. With a range of reflector shapes, a particular family of luminaires can provide uniform lighting free of streaks and hot spots in a range of beam spreads from very narrow to very wide.

REFLECTOR SURFACES

Reflectors can have smooth, highly polished surfaces, or their surfaces can be shaped or treated in some way to give slight diffusion. Light reflected from a smooth, highly polished surface follows the law of regular reflection. If a clear arc source is used with such a reflector, small changes in lamp position can produce significant changes in light distribution. Changes in lamp position can produce streaks of irregular shapes, causing rings or swirls on the surface that is being lighted.

In addition, a smooth, highly polished reflector surface can reflect the image of the light back onto the arc tube, changing the lamp's operating characteristics. High-pressure sodium (HPS) sources, for instance, will have reduced life in this situation.

For these reasons, some diffusion is usually added to the reflector surface. This scatters light so that the beam pattern is smoother. It also prevents light from being redirected onto the arc tube.

Diffusion can be added to the reflector surface by etching or by treating or coating the surface in some way. The surface can be grooved, faceted, or fluted to add diffusion. Some reflectors are formed from myriads of tiny prisms. Others are painted, or one type of material is bonded to another.

Each of these approaches results in a different pattern of light output. In addition, there are variations in products from one manufacturer to another as well as within one manufacturer's line. Luminaire efficiency can vary as much as 10 to 15 percent with the same reflector design and material.

The designer should always request specific product information before deciding which luminaire to use. This information should include graphs and tables of photometric performance.

SHIELDING ANGLE AND BRIGHTNESS CONTROL

The *shielding angle* is the angle between the edge of the reflector and the bottom of the light source, which is the bottom of the bulb for phosphor-coated or etched bulbs and the bottom of the arc tube for clear lamps. The *flashed* part of the reflector is that part where the reflected brightness of the lamp is visible. If the flashed part is at or above the shielding angle, then brightness on the inside of the reflector will appear low when viewed above the shielding angle, and the luminaire will be comfortable to view.

BEAM SHAPE

The beam pattern from a luminaire can be either symmetrical or asymmetrical. This is controlled by reflector and/or refractor design. A symmetrical beam pattern is one in which the same amount of light is emitted at each vertical angle around the nadir, or 0 degrees, so that the light forms a circular pattern.

A beam pattern can also be asymmetrical, which means that the light forms a shape other than a circle. In most cases the shape is elliptical and the beam is symmetrical about the 0- and 90-degree horizontal axis. In other cases the beam can have almost any shape but is usually symmetrical left to right of the 0- to 180-degree horizontal axis.

Reflector Performance

This discussion of reflector performance concerns reflectors which are open at the bottom or are covered with clear glass. Diffusers or refractors change reflector performance.

CIRCULAR REFLECTORS

Circular reflectors are designed as if the light source were a point source. The light distribution of the entire reflector is determined by the sum of the reflections produced from each portion of the reflector surface.

An HID source is not a point source. It is a long tube having different brightnesses both across the tube and along its length. The reflector design centers around a single point in the arc tube but must take into account the contribution from the rest of the lamp. The larger the arc tube or light source, the harder it is to produce a precise beam shape.

There is no way for a point on the reflector to concentrate the light from the lamp into a beam narrower than the angle produced by the edges of the light source. Increasing the size of the light source will spread the light pattern and reduce the ability to produce either a precise beam or one with a narrow distribution. Etching the lamp or coating it with phosphor will substantially enlarge the source size and make it impossible to produce anything but a wide-beam reflector. This does not reduce the luminaire efficiency. It may, however, reduce the utilization if long projection distances are involved.

FLASHED AREA

The narrower the beam and smaller the light source, the more restricted will be the angle where the reflector appears flashed. The reflector of a large searchlight will appear flashed only if a viewer looks right into the searchlight. Otherwise the reflector appears dark.

As the source size is increased or diffusion is added to a reflector, the reflector will appear flashed regardless of the angle at which it is viewed. If the beam spread of a clear lamp and polished reflector is wide, then diffusing the light source or reflector will make the beam narrower. If the beam spread is narrow and diffusion is added, the beam will get wider.

PERFECT DIFFUSION

A completely diffuse lamp and reflector will approach a circular distribution, as shown in Fig. 3-4. (The direct contribution of the lamp could change this distribution somewhat.) The luminous intensity of such a luminaire approaches a cosine curve, which obeys Lambert's law. The formula to express Lambert's law is

$$I = I_{\max} \cos \theta$$

where I = candlepower, or luminous intensity, expressed in candelas.

The cosine of the angle θ is 1 at 0 degrees, or straight in front of the luminaire, and 0 at 90 degrees. Thus candlepower is at its maximum at 0 degrees and decreases to 0 at 90 degrees.

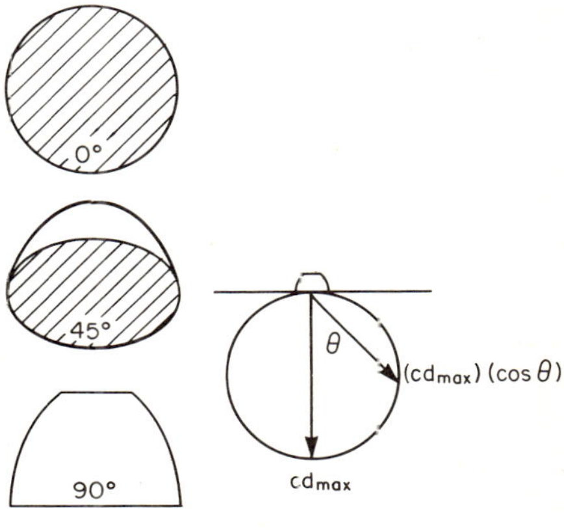

Fig. 3-4 A perfectly diffuse surface will reflect light in a cosine curve, which approaches a circular distribution. The reflector will appear flashed at any viewing angle.

A white, painted matte finish provides the most diffuse reflector surface. With a perfectly diffusing reflector surface, luminous intensity is spread out, but the reflector exercises almost no control of light output. In contrast, a point source used with a highly polished reflector is capable of producing very high luminous intensity in a given direction, but at some angles there will be no luminous intensity at all. Most reflector designs fall somewhere between these extremes.

BRIGHTNESS

The term *brightness* can have more than one meaning. In its more favorable sense, the term can refer to the luminous intensity of a luminaire, since this is a function of the brightness of the flashed area of the reflector or refractor plus whatever is contributed by the lamp directly. The brightness of the flashed area is proportional to the brightness of the lamp and the finish of the reflector. This brightness is difficult to calculate but can be measured; it is always less than lamp brightness.

On the other hand, *brightness* can be a subjective term referring to how comfortable a person feels viewing a given luminiare (i.e., whether the luminaire is "too bright"). This kind of brightness is theoretically determined by the brightness of the luminaire at a given angle divided by the *projected area* of the reflector, that is, the area of the reflector that is seen from a certain angle. However, this calculation gives only average brightness. Whether or not a luminaire appears too bright is a function of both peak brightness and average brightness. A highly polished reflector and a faceted one of the same size could have the same average brightness, but one might appear comfortable to view and the other might not, because of peak brightness.

Spreading the flashed area of a reflector will reduce peak brightness, and up to a point this will reduce the possibility of glare, or of the luminaire appearing too bright. Evaluating the visual brightness or glare of a luminaire is best done subjectively or by comparison with a luminaire that has already been evaluated.

TROUGH-TYPE REFLECTORS

Trough-type reflectors are used where the long dimension of the arc runs perpendicular to the major beam. These reflectors have inherently less control than circular reflectors, because 25 percent or more of the lamp's output does not strike the reflector at all. The reflector controls the output of the lamp predominantly in a line perpendicular to the long dimension of the arc tube.

Trough reflectors are used where control is necessary in only one plane, either horizontal or vertical. A narrow beam distribution is possible in only one plane perpendicular to the long axis of the arc tube. The distribution in the other direction will usually be medium to wide. The entire beam shape is thus asymmetric.

If clear arc sources are used in these reflectors, then small changes in lamp position will result in dramatic changes in beam spread because of the proximity of the lamp to the reflector surface. The position of the light source within the reflector is much more critical.

An advantage of using a trough-type reflector is that the arc tube or lamp can be put close to the outside edge of the reflector. It is thus possible to flash the entire lower edge of the reflector, producing high candlepower at a very wide distribution. This is the reason behind the horizontal lamp position in roadway luminaires.

Glass Covers

Luminaires can be open on the bottom, or they can be covered with a flat, clear glass or a diffuser or refractor. A clear cover glass provides no optical control. It is used to protect the lamp, keep the inside of the luminaire clean, and enhance appearance.

Of the light incident on a flat glass surface, approximately 4 percent is reflected off the first, air-to-glass surface and the same percentage off the second, glass-to-air surface. Depending on the quality of the glass, 2 to 8 percent of the light is absorbed by the glass. The use of a cover glass can thus be expected to reduce the efficiency of a luminaire approximately 10 percent.

The reflection of light off a glass surface is not constant. Above an incident angle of 50 degrees reflection increases rapidly. At an incident angle of 75 degrees it is approximately 30 percent. This effect is most apparent in wide-angle roadway and high-mast luminaires, which require high candlepower at 60 to 75 degrees. The high-angle reflection off the glass surface will tend to reduce the high-angle candlepower. Using a dished cover glass or a pebbled surface will reduce this problem.

Refractors

Refractor design is based on the law of refraction; light is bent into a specific pattern, either with some sort of lens or with prisms. The entire control of light output can be accomplished with a refractor. This is typical of many low-wattage area and roadway luminaires. However, the control possible with refractors is limited to the maximum bending that can be achieved with an individual prism plus the interreflectance which different prisms make with one another.

Luminaire efficiency can be high when a refractor is used alone, but utilization will not be so high as with a reflector-refractor combination.

One major use of refractors is with roadway-type luminaires, where prisms on the side of the refractor are used to elevate the maximum candlepower from a luminaire by 5 to 10 degrees over what would be possible with a reflector alone. Refractors can also be used to create an asymmetrical beam from a symmetrical reflector.

Refractors can be formed from glass, acrylic, or polycarbonate plastic. Polycarbonate has unusually high strength and is a good choice where vandalism is a problem. Polycarbonate tends to yellow and become brittle after prolonged exposure to ultraviolet rays, but it can be treated so it is more resistant to this effect. Glass and acrylic both have excellent optical properties and stability.

Diffusers

A *diffuser* is any translucent material which is used to hide a lamp. Diffusers are made of glass, plastic, or polycarbonate. The addition of pigments to a refractor can add some diffusion qualities, or a diffuser can be used below a lamp in place of a refractor.

Diffusers can be any shape. Their purpose is to spread the brightness of a light source over a larger area, so that peak brightness is minimized. Diffusers can be used where brightness control is required but there is no need for precise photometric control.

Luminaires with diffusers have a lower efficiency than open or enclosed reflectors or reflector-refractor combinations.

Louvers and Visors

Louvers and visors are physical barriers to light output. They are usually painted black so that they absorb light readily. Louvers and visors are used on outdoor luminaires to block light output that might be glaring and objectionable.

Visors provide limited control of luminaire brightness when they are located on the sides of the top of the luminaire, as shown in Fig. 3-3. The effectiveness of the visors is a function of their length. As a result, they are useful only in controlling the wide-angle brightness of a luminaire. For top visors, this wide-angle brightness is normally spill light, so there is very little loss in the efficiency or utilization of the luminaire.

As brightness control gets further into the main beam, however, visor length must be increased to the point where visors are impractical. In that case, a louver inside or in front of the luminaire would have to be used.

Louvers can absorb as much as 50 percent of the total light output of a luminaire, so they should be used only if absolutely essential. Proper location of floodlights or the use of trees or fences to block stray light is a more efficient approach.

If louvers are used, the lighting system should be designed with the louvers in place. Louvers will totally change the photometric performance of the luminaire, and the designer should request the altered photometric distribution from manufacturers.

Seals and Filters

A luminaire used outdoors and in dirty indoor applications should be sealed to prevent dirt from accumulating on the reflector. The exterior surface of the door glass will still get dirty but is easier to clean than the entire optical assembly.

Sealed luminaires become partially pressurized when the lamp is turned on. For a sealed luminaire to work correctly the pressure must be contained by latches and gaskets. With a large gasket area and spring-type latches this will not occur. The expanded air will be driven out of the luminaire through the path of least resistance.

When the luminaire is turned off, the process will be reversed, and outside air will be drawn into the luminaire. The luminaire is said to breathe. If there are any contaminants, they will settle out on the reflector or inside surface of the door glass. This can create a worse situation than if the luminaire were left open.

A solution is to provide a low-pass filtered passageway into the luminaire. Most solid contaminants, such as dirt and dust, can be kept out of optical assemblies with gaskets and various types of filters. Adsorptive filters of activated charcoal keep out gaseous and molecular contaminants, as well as dust and dirt.

Quartz Standby Lighting

Luminaires can be provided with automatically switched quartz standby lighting where momentary power interruptions or extreme voltage dips can extinguish an HID lamp. A single-ended quartz lamp is placed in the same reflector with an HID lamp, as shown in Fig. 3-5. These supplementary sources are placed not in every luminaire but in key spots throughout an area in about 10 percent of the luminaires. After a power interruption or when a system is first switched on, the quartz lamp will light and remain on until the HID lamp reaches about 60 percent of light output.

Wiring for the quartz lamp is internal to the ballast, and therefore the quartz lamp voltage is independent of the lighting system voltage. When sizing branch circuits, note that the luminaire will draw additional current when the quartz and HID lamps are operated simultaneously.

Quartz lamps can also be wired separately as part of an emergency lighting system. They would then be energized by an emergency generator or some other emergency power supply.

These supplementary systems are needed for safety. They are especially important when mercury or metal halide sources are being used, because restrike time can be as much as 15 minutes.

Modifications for Special Environments

Some luminaires can be ordered with factory modifications that make them suitable for use in special environments. The designer should know

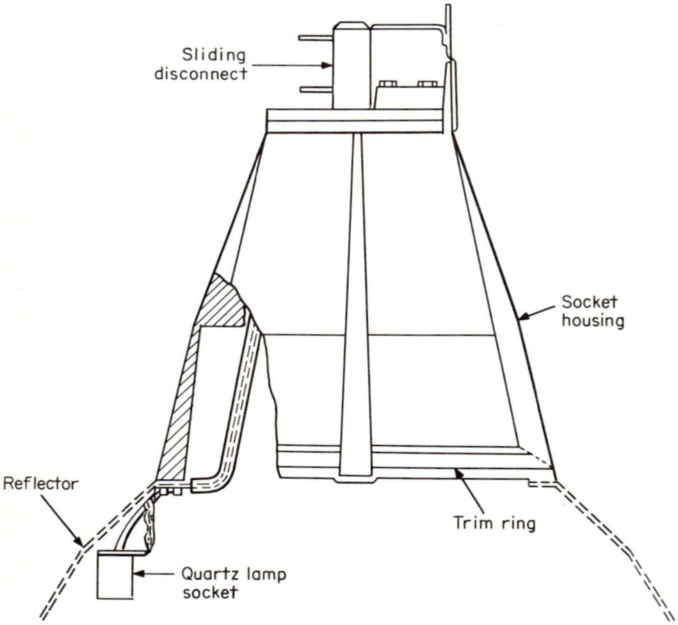

Fig. 3-5 Quartz lamps can be mounted in selected luminaires to serve as standby lighting. The lamps can be wired with the luminaire or can be on a separate emergency circuit.

as much as possible about expected ambient conditions before choosing luminaires.

CORROSIVE ENVIRONMENTS

If a luminaire is to be used in a corrosive environment, all exposed screws, rivets, or die-cast parts should be made of corrosion-resistant materials or painted with special industrial paints. Reflectors may need special protection as well.

WET, DAMP, OR DUSTY ENVIRONMENTS

Luminaires to be used in wet or damp locations should bear an Underwriters Laboratories label stating that they are suitable for such applications. These luminaires have a special gasketed construction. Gaskets are provided at all points where water entry is probable. These gaskets will keep out dust as well as water, so these luminaires can be used in dust-laden environments as well.

SEVERE-DUTY APPLICATIONS

The techniques used to prepare luminaires for corrosive environments and wet and dusty environments are combined for use in wet, dusty, and corrosive atmospheres. Corrosion-resistant materials and finishes are used, and gasketing is placed at all points where water or dust entry is probable.

HAZARDOUS LOCATIONS

Hazardous locations are those in which there is danger of fire or explosion because of the presence of flammable gases or vapors, combustible dust, or easily ignitable fibers or flyings. The National Fire Protection Association publishes a *National Electrical Code* which defines and classifies hazardous locations according to class, division, and group.

The code contains performance guidelines for equipment to be used in hazardous locations. No surface of the equipment should exceed the ignition temperature of the gas, vapor, dust, or flyings which are present, and there should be no danger of sparks. Maximum safe temperatures for various classifications are included as part of the code. These temperatures may restrict the lamp wattage that can be used in a particular environment.

Underwriters Laboratories tests luminaires for use in hazardous locations and lists them as "UL 844 approved" if they pass the rigorous requirements. A luminaire so listed will bear a label saying that it is listed for use in a specific class of hazardous locations.

A designer choosing lighting for a hazardous location should be familiar with both the *National Electrical Code,* Chap. 5, "Special Occupancies," and the Underwriters Laboratories publication, *Standard for Safety, Electric Lighting Fixtures for Use in Hazardous Locations,* UL 844.

Corrosion Resistance of Materials

Materials used in luminaires for industrial applications will sometimes be subjected to chemical environments. Some materials have excellent resistance to some environments but will quickly break down when subjected to others. Designers should be aware of the general nature of materials and their resistance characteristics.

The following discussion is intended only as a guide; the designer is simply cautioned to investigate further if certain materials are to be used in potentially harmful environments. Manufacturers should be consulted, since some materials are treated to make them more corrosion-resistant and others are not.

A luminaire's resistance to corrosive environments will change with the concentration of the corrosive material, the temperature, and the amount of moisture in the air. How often the luminaire will be subjected to the material is another factor. as well as whether the luminaire will be washed between exposures. Any coating or finishing process which the material has received may also make a difference.

Table 3-1 gives resistance ratings for certain chemical environments for various reflector and refractor materials and for two types of paint. Ratings are for corrosion at room temperature. Corrosion resistance is rated as **E** (excellent), **G** (good), **F** (fair), and **P** (poor). These ratings stand for the following effects:

E = essentially no effect

G = some slight effect, but material will probably last the life of the unit

F = moderate effect that might cause problems during the life of the unit

P = severe attack that will significantly shorten the useful life of the unit

The reflector materials in Table 3-1 include aluminum, Alzak aluminum (developed by the Aluminum Company of America), and glass bonded to aluminum. The data for aluminum also apply to aluminum castings.

Table 3-1 also gives the resistance properties of glass, acrylic, and polycarbonate refractors to the stated chemicals, and similar information for acrylic paint and epoxy paint. These paints have good to excellent resistance to many chemicals and are often used on luminaires specifically designated for corrosive environments.

Luminaire Dirt Depreciation

One of the most important factors in luminaire choice and lighting design is luminaire dirt depreciation (LDD). It is also one of the hardest things to determine with any accuracy.

Luminaire dirt depreciation reflects loss of light output for a particular luminaire under certain ambient dirt conditions. Figures for LDD are given by manufacturers, but they are invariably optimistic. They assume a regular maintenance program, but most users are lax about cleaning, relamping, etc.

Even if an industry is careful about maintenance, it is still hard to estimate a figure for LDD. Depreciation within a plant is not consistent. Variations in industrial procedures, air currents, etc., can result in different LDD from one bay to another or even within the same bay.

Nevertheless, the designer must estimate a figure for LDD in order to design for maintained light levels rather than initial levels. The LDD is one of the main components of the light-loss factor. Another is lamp lumen depreciation (LLD), which takes into account loss in lumen output as a lamp ages.

In order to estimate LDD a designer should study the design of the particular luminaire to be used and the ambient dirt conditions where the luminaire will be installed. Experience within the same plant or with a luminaire of a similar type can be used as a guide. Lighting maintenance firms can also be consulted.

LUMINAIRE DESIGN

High-intensity discharge luminaires can be roughly divided into four categories widely used in industry. Some types of designs are more susceptible to dirt depreciation than others.

Because open-bottom luminaires have no barrier on the bottom, they tend to accumulate dirt inside, directly on the lamp and reflector. Luminaires with open bottoms and open tops have vents in the top of the reflector. As air is heated by the light source it rises and passes out through the vents, carrying with it some of the dirt present in the air.

Enclosed and gasketed luminaires are tightly sealed so that most of the dirt in the air cannot enter the optical assembly. Dirt will accumulate only on the bottom of the lens, diffuser, or refractor.

Table 3-1 Corrosion resistance of materials

Material	Resistance to: Acids*	Bases†	Salts‡	Wet Gases§	Solvents	Oils and Fuels
Reflector Materials						
Aluminum	P: Formic Hydrochloric Hydrofluoric Sulfuric F-E: All others	G: Ammonium hydroxide P: Potassium or Sodium hydroxide	P: Ammonium chloride Copper sulfate Ferric chloride Magnesium chloride Mercuric chloride Nickel chloride Nickel sulfate Potassium chloride Sodium hydrochlorite Sodium phosphate Zinc chloride Calcium chloride F-E: All others	P: Chlorine F-E: All others	G-E	G-E
Alzak§ aluminum	G: Fatty P: Chromic Hydrochloric Hydrofluoric Nitric Phosphoric Sulfuric F: All others	F: Ammonium hydroxide P: Potassium or Sodium hydroxide	F	G: Carbon dioxide E: Carbon disulfide F: All others	E	E
Glass bonded to aluminum	P: Hydrofluoric F-G: Chromic G-E: All others	G: Ammonium hydroxide F: Potassium or Sodium hydroxide	G: Borax G-E: Some sodium salts E: All others	E	E	E
Refractor Materials						
Glass	F-P: Hydrofluoric E: All others	F: Potassium or Sodium hydroxide E: Ammonium hydroxide	E	E	E	E

Acrylic	G: Acetic Formic Hydrochloric Picric E: Boric Citric Fatty F: All others	G	E	F: Chlorine G-E: All others	P-F	P: Hydraulic fluid (ester) G: All others
Polycarbonate	G: Acetic Formic Hydrochloric Picric E: Boric Citric Fatty F: All others	P	F: Some sodium salts E: All others	P: Ammonia F: Chlorine G-E: All others	P-F	P: Hydraulic fluid (ester) G: All others
			Paints			
Acrylic	G: Boric Citric Hydrochloric F: All others	G: Ammonium hydroxide F: Potassium or Sodium hydroxide	G	G-E	P: Methylene chloride F: Methyl ethylketone Perchloro-ethylene Trichloro-ethylene G-E: All others	E
Epoxy	G: Fatty Hydrochloric Nitric Phosphoric Picric Sulfuric E: Boric Citric F: All others	G: Ammonium hydroxide E: Potassium or Sodium hydroxide	E	E	G-E	E

*Acids are all concentrated with enough water to cause ionization.

†All bases and salts are nonvolatile materials and must be present as a mist to cause detrimental effects, except for ammonium hydroxide and decomposition of some ammonium salts.

‡Dry gases will be less harmful.

§Proprietary term of Aluminum Company of America.

Some contaminants can enter sealed luminaires as air is forced in and out as the luminaire cools and heats up again. The filter used in sealed and filtered luminaires filters out any dirt and gaseous matter that is breathed into the optical assembly.

AMBIENT DIRT CONDITIONS

The ambient dirt conditions in which a luminaire will be mounted must be known before an LDD factor is assigned to that luminaire. Five degrees of dirt conditions are shown in Table 3-2. These are classifications of the Illuminating Engineering Society of North America (IES).

Ambient Temperature Conditions

All lighting equipment should be used under the ambient temperature conditions for which it was designed. If a luminaire is to be placed in an excessively hot or excessively cold environment, care should be taken to choose components which will operate effectively under those conditions. Starting requirement for HID sources may be affected by temperature; ballasts rated for the expected temperature should therefore be used. Under extreme conditions, ballasts must be remotely mounted.

Even if lamps will start, damage may be done to other parts in extreme temperature conditions. The first thing to fail is usually the capacitor. This is the component most sensitive to over- or under-temperature operation.

Some plastic components may have temperature limits beyond which they will soften or deteriorate. Suppliers should be consulted before final choices of materials are made.

Vibration

All luminaires are subject to some vibration, but mounting them on moving cranes can increase vibration substantially. Mild vibration will have no adverse effect on ballast or lamp life, but severe vibration will cause failure of the lamp or arc-tube support structure or of the outer envelope at the location of the socket. Severe vibration could also cause fatiguing of the cast ballast enclosure so that it will eventually fail.

If luminaires are to be mounted in positions where they will be subject to extreme vibration or shock, special brackets should be used. These hold the ballast remote from the reflector assembly and reduce strain on the connections. The ballast can also be mounted on a flexible connector, and the optical assembly on a resiliant structure.

Table 3-2 Five degrees of dirt conditions

	Very Clean	Clean	Medium	Dirty	Very Dirty
Generated dirt	None	Very Little	Noticeable but not heavy	Accumulates rapidly	Constant accumulation
Ambient dirt	None (or none enters area)	Some (almost none enters)	Some enters area	Large amount enters area	Almost none excluded
Removal or filtration	Excellent	Better than average	Poorer than average	Only fans or blowers if any	None
Adhesion	None	Slight	Enough to be visible after some months	High—probably due to oil, humidity, or static	High
Examples	High-grade offices, not near production; laboratories; clean rooms	Offices in older buildings or near production; light assembly; inspection	Mill offices; paper processing; light machining	Heat treating; high-speed printing; rubber processing	Similar to Dirty but luminaires within immediate area of contamination

SOURCE: *IES Lighting Handbook,* 5th ed., Illuminating Engineering Society of North America, New York, 1972.

Mounting brackets for standard floodlights are designed for fixed locations and can fail if not properly supported. When floodlights are used on portable equipment such as traveling cranes, manufacturers should be consulted for advice on luminaire strength-mounting techniques.

Poles should not be mounted on a base that is subject to vibration or shock, such as on a bridge. This is especially true of roadway lights mounted on 1- to 2-meter arms; the motion of the base can be amplified by the crossarm, causing failure in the support mechanism.

Lighting System Noise Criterion (LS-NC) Ratings

Lamp-ballast-luminaire combinations give off a slight hum. This will not usually be a problem unless the area in which the luminaires are used is so quiet that the hum becomes noticeable. There may be other noise in the area, such as that coming from air-conditioning and ventilation equipment, people, music, typewriters, machines, cash registers, etc. In that case, a slight hum from the luminaires would never be noticed. However, if the hum is clearly audible over the ambient noise, it could be objectionable.

The National Electrical Manufacturers Association (NEMA) publishes a standard that is used to determine in advance whether the noise from a specific lighting system will be annoying if used in a particular situation. The standard is *H-I-D Lighting System Noise Criterion (LS-NC) Ratings,* Standards Pub. No. LE 2-1974. The first section gives procedures to follow in order to assign an LS-NC rating to a lamp-ballast-luminaire combination. The rating is for a level of 1076 lux (100 footcandles) average maintained in an indoor installation having acoustically hard walls and floors and an acoustic tile ceiling. Some manufacturers specify certain spacings when assigning an LS-NC rating.

The second section shows how to apply LS-NC ratings. It includes correction factors for other light levels, spacings, or room surfaces. The section also gives ways of comparing corrected LS-NC ratings with expected or measured ambient sound levels to decide if the lighting system will be audible.

Inspecting a lighting installation before machinery is turned on, offices are occupied, or air-conditioning and heating equipment is in use can give an unfair picture of how annoying the noise from the lighting system will be. In a completely empty, quiet environment the hum may be noticeable, but it may be entirely masked by ambient noise as soon as the area is in use.

DETERMINING CORRECTED LS-NC RATINGS

Whether or not a system will be audible can be determined by the following procedure:

1. From product literature, obtain the LS-NC rating for the proposed lamp-ballast-luminaire choice. This is given in the form "NC-42," "NC-45," etc. This rating is sometimes for a particular spacing.

For example, assume that the lighting unit is rated NC-42. This is a decibel (dB) rating; if the room decibel rating is the same or higher, the system will probably be inaudible. The A-weighted sound pressure level—dBA—that would mask the LS-NC rating of 42 is the LS-NC rating plus 8, or 50 dBA. Differences of 5 or less in LS-NC ratings of different systems are not significant.

2. Apply correction factors if they are applicable, for light levels other than 1076 lux (100 footcandles) or spacing other than that specified by the manufacturer. The appropriate correction factor is known as L_E and can be read from Fig. 3-6.

Assume that the expected level of illumination E will be 50 footcandles, or 500 lux, instead of 100 footcandles. The correction factor is −3, and the corrected LS-NC rating is therefore 42 − 3 = 39.

If the reference spacing S_0 used for testing is the same as the actual spacing S, then the ratio S/S_0 is 1.0 and the correction factor is 0. Assume, though, that the actual spacing is 1.4 times the reference spacing. The correction factor is then −3, and the corrected LS-NC rating is therefore 39 − 3 = 36.

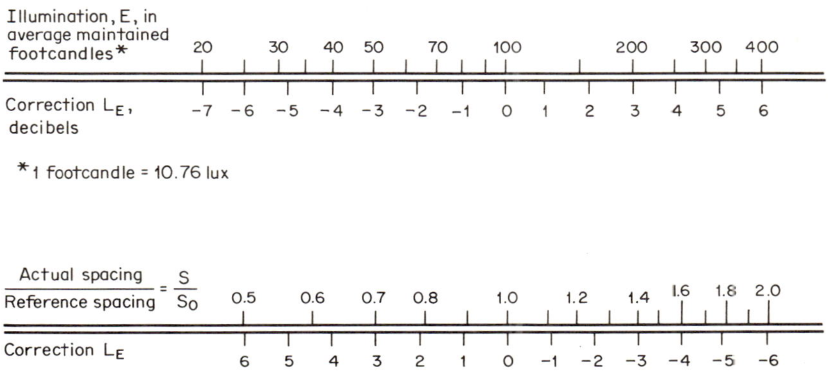

Fig. 3-6 Correction factors for light levels other than 1076 lux (100 footcandles) and spacings other than the test spacing. (This data is reprinted with permission from *H-I-D Lighting System Noise Criterion (LS-NC) Ratings,* Standards Pub. No. LE 2-1974, National Electrical Manufacturers Association, New York, 1974.)

3. Apply correction factors for room reverberation levels other than those used in the test room. Acoustical correction factors are given in Table 3-3.

A live room would be an area such as a gymnasium or swimming pool. Most factories would be classified as medium-live rooms to which a correction factor of +6dB would apply. The corrected LS-NC rating would then be 36 + 6 = 42. The final corrected rating for this lamp-ballast-luminaire is NC-42.

FINDING THE NOISE CRITERION LEVEL OF THE ROOM AMBIENT

The noise criterion level in the space where the luminaires will be installed should be measured or assumed with the space occupied and heating or air-conditioning systems on. The noise criterion level can be estimated by comparing noise surveys of installations similar to the one being planned, by using the noise criterion design level used to specify the air-conditioning system, or by reading the estimated ambient sound levels from Table 3-4. If the room is already occupied, a sound pressure level can be measured. Note that this level will be higher by 8 than the noise criterion decibel level.

In industrial areas, except for offices, expected noise criterion ratings

Table 3-3 Acoustical correction factors

Room Environment	Description	Corrections
Medium-dead room	All room surfaces have good sound absorption properties, e.g.,	−3 dB
	Floor . . . . carpeted	
	Ceiling . . . . . noise reduction coefficient (NRC) 0.70 or higher	
	Walls . . . . draperies	
	Furniture . . . . upholstered	
Average room	Some room surfaces are sound absorptive, e.g.,	0 dB
	Floor . . . . tiled or wood	
	Ceiling . . . . . noise reduction coefficient (NRC) 0.60 or higher	
	Walls . . . . sound reflective	
	Furniture . . . . sound reflective	
Medium-live room	All room surfaces are reasonably hard, e.g.,	+6 dB
	Floor . . . . tiled or wood	
	Ceiling . . . . plastered or gypsum board	
	Walls . . . . wood paneling, gypsum board unpainted brick, cement block, plaster	
Live room	All room surfaces are hard, e.g.,	+10 dB
	Floor . . . . concrete	
	Ceiling . . . . plastered gypsum or concrete	
	Walls . . . . painted brick or concrete	

SOURCE: Reproduced with permission from *H-I-D Lighting System Noise Criterion (LS-NC) Ratings,* Standards Pub. No. LE 2-1974, National Electrical Manufacturers Association, New York, 1974.

Table 3-4 Range of design levels for air-conditioning and ventilation system sound control

Type of Area	Noise Criterion Level
MANUFACTURING AREAS	
Foreman's office	40–50
Assembly lines, light machinery	45–70
Foundries, heavy machinery	55–75
OFFICES	
Board room	20–30
Conference rooms	25–35
Executive office, supervisor office, reception room	30–45
General open offices, drafting rooms	35–50
Halls and corridors	35–55
Tabulation and computation	40–60

SOURCE: Reproduced with permission from *H-I-D Lighting System Noise Criterion (LS-NC) Ratings,* Standards Pub. No. LE 2-1974, National Electrical Manufacturers Association, New York, 1974.

are from 45 to 75. This level is higher than the corrected LS-NC rating of 42 for the proposed luminaire choice and so the lighting system should be inaudible.

USING AN AUDIBILITY CHART

Figure 3-7 is an audibility chart which can be used to estimate the audibility of a lighting system. The corrected LS-NC level can be entered on the chart as a vertical line. The estimated noise criterion level of the room ambient in dB can be found on the right side, and the measured room ambient sound level in dBA can be found on the left side. A line from either of these points is then drawn directly across the chart. If it intersects in the inaubible area with the line representing the LS-NC rating, the system will not be heard above the ambient noise.

Entered on the graph are the lines for a lighting system with an LS-NC rating of 42 and a room with noise criterion level of 42 and a dBA rating of 50. In this situation, the lighting system is at the threshold of audibility and should be inaudible. Any room noise criterion level above 42 will also mask the system noise.

SUMMARY

The following summary is reprinted with permission from NEMA Pub. No. LE 2-1974:

1. For a lighting system whose corrected LS-NC rating is equal to or less than the noise criterion (NC) level specified or expected in

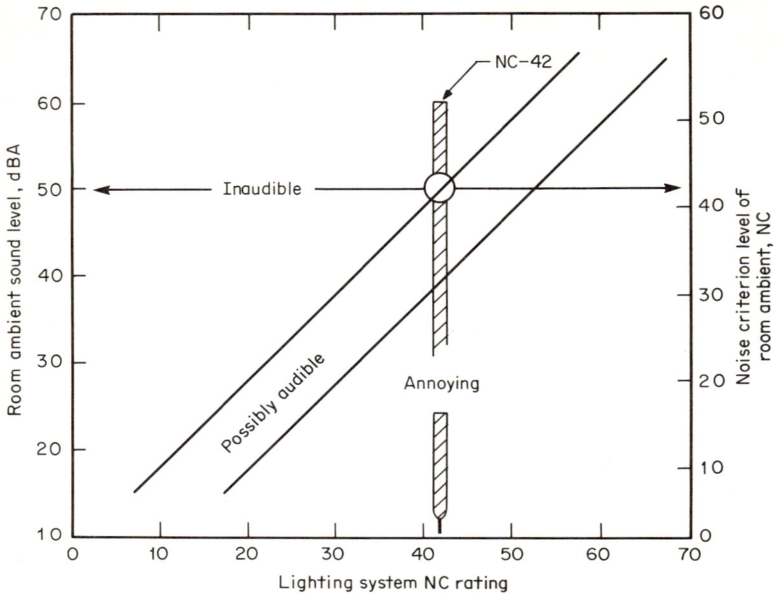

Fig. 3-7 A lighting system will be inaudible if the noise criterion level of the room ambient is the same as or greater than the LS-NC rating of the system or if the dBA level of the room is higher by 8. (This graph is reproduced with permission from Standards Pub. No. LE 2-1974, National Electrical Manufacturers Association, New York, 1974.)

the space to be occupied, hum from the lighting installation will probably be inaudible.

2. For a lighting system whose corrected LS-NC rating is between 0 and 10 decibels higher than that specified or expected from the air-conditioning equipment, the hum may be audible to some occupants.

3. For a lighting system whose corrected LS-NC rating is more than 10 decibels above the ambient noise level, the occupants may volunteer complaints.

Bibliography

Barnes, J. T.: "Plastics—Pros and Cons for Outdoor Lighting," *Lighting Design and Application,* December 1972, pp. 29–31.

Burkhardt, W. C.: "High-impact Acrylic for Lenses and Diffusers," *Lighting Design and Application,* April 1977, pp. 12–20.

Burkhardt, William C.: "Acrylics and HID Light Sources," *Journal of the Illuminating Engineering Society,* July 1977, pp. 224–232.

Clark, Francis: "Accurate Maintenance Factors—Part Two (Luminaire Dirt Depreciation)," *Illuminating Engineering,* January 1966, pp. 37–46.

"Comparative Corrosion Resistance of Selected Metals and Non-Metals," *Metal Progress,* February 1977.

"Hazardous (Classified) Locations," *National Electrical Code,* Article 500, National Fire Protection Association, Boston, 1978.

H-I-D Lighting System Noise Criterion (LS-NC) Ratings, Standards Pub. No. LE 2-1974, National Electrical Manufacturers Association, New York, 1974.

"How Materials Stand Up to Corrosion and Chemical Attack," *Material Engineering,* February 1971.

"Light Control and Luminaire Design," and "Lighting Calculations," *IES Lighting Handbook,* 5th ed., secs. 6 and 9, Illuminating Engineering Society of North America, New York, 1972.

Manual for High-Intensity-Discharge (H-I-D) Lamps, Standards Pub. No. LE 3-1974, National Electrical Manufacturers Association, New York, 1974.

"Materials Selector 78," *Materials Engineering,* Reinhold Publishing Company, Stanford, Conn., 1978.

Modern Plastics Encyclopedia, 1977–1978.

Palko, Ed: "A Plant Engineer's Guide to HID Luminaires," *Plant Engineering,* Aug. 8, 1974, pp. 58–63.

Sherwood, M.: "Aluminum Reflectors: Improved Characteristics," *Lighting Design and Application,* December 1972, p. 34.

Staley, Karl A.: "Light Control," *Fundamentals of Light and Lighting,* chap. V, August 1960.

Standard for Safety, Electric Lighting Fixtures, UL 57, 12th ed., Underwriters Laboratories, Inc., Northbrook, Ill., Sept. 30, 1977.

Standard for Safety, Electrical Lighting Fixtures for Use in Hazardous Locations, UL 844, 7th ed., ANSI C 33.28-1972, Underwriters Laboratories, Inc., Northbrook, Ill., June 1, 1972.

Weil, Luis G.: "Picking a Plastic," *Lighting Design and Application,* December 1973, pp. 37–39.

4

Interior Lighting Design

As in all areas of lighting design, interior design requires not only a basic knowledge of the equipment available but also an understanding of design procedures. This in turn means that the lighting practitioner must be familiar with photometric data that is published by manufacturers for their products.

This chapter will describe how to read and use indoor photometric data for high-intensity discharge (HID) luminaires. Lighting calculation procedures are given which can be used to obtain average levels of illumination. Other information helpful in selecting sources and luminaires is also included. The next chapter, Chap. 5, uses this information to solve a typical industrial lighting design problem and analyzes the results.

Chapter 7 includes charts which can be used to calculate illumination levels at any point on a horizontal or vertical surface.

Photometric Data

The Illuminating Engineering Society of North America (IES) specifies various methods of photometric testing of luminaires. The results of these tests are published by manufacturers in the form of photometric data, such as that shown in Fig. 4-1. This data is for a 400-watt high-pressure sodium (HPS) luminaire designed to be used at low mounting heights. Photometric data can be used to compare the performance of similar luminaires, although choices should not be made on the basis of slight differences.

CANDELA DISTRIBUTION GRAPH

When a luminaire is tested, it is placed at the center of an imaginary sphere having a radius equal to the test distance. The test distance should be 5 to 10 times the maximum diameter or dimension of the luminaire so that the luminaire appears as a point source. Measurements of light intensity in candelas are made from every direction. Circular (symmetrical) reflectors are sometimes rotated to obtain the average intensity at each vertical angle. The candela measurements are plotted on polar coordinates to show the shape of the candela distribution curve. This shows luminous intensity in any direction relative to the conventional position of the luminaire in service.

Figure 4-1 includes a graph of the candela distribution for a sample luminaire. The luminaire was tested at a distance of 7.6 meters (25 feet).

Test data is based on a lamp rated at 1000 lumens. Candela distribution charts were formerly drawn on the basis of rated initial lamp lumens, but most newer data is based on 1000 lumens. This makes it easier to compare

GENERAL (GE) ELECTRIC

PHOTOMETRIC DATA

COEFFICIENTS OF UTILIZATION - ZONAL CAVITY METHOD

EFFECTIVE FLOOR CAVITY REFLECTANCE 20 PER (Pfc = 0.20)

	EFFECTIVE CEILING CAVITY REFLECTANCE - PERCENT														
	80			70			50			30			10			0
% WALL REFLECTANCE / ROOM CAVITY RATIO (RCR)	50	30	10	50	30	10	50	30	10	50	30	10	50	30	10	0
0	.83	.80	.78	.80	.79	.77	.76	.75	.74	.72	.72	.71	.69	.68	.68	.66
1	.74	.71	.68	.72	.69	.67	.68	.66	.64	.65	.63	.62	.62	.61	.59	.57
2	.65	.61	.57	.63	.59	.56	.60	.57	.54	.58	.55	.53	.55	.53	.51	.49
3	.57	.52	.48	.56	.51	.47	.53	.49	.46	.51	.47	.44	.49	.46	.43	.41
4	.50	.44	.40	.49	.43	.39	.47	.42	.38	.45	.41	.37	.43	.39	.36	.35
5	.44	.37	.33	.43	.37	.32	.41	.36	.32	.39	.35	.31	.38	.34	.30	.29
6	.38	.32	.27	.38	.31	.27	.36	.31	.26	.35	.30	.26	.33	.29	.25	.24
7	.34	.27	.22	.33	.27	.22	.32	.26	.22	.31	.25	.22	.29	.25	.21	.20
8	.30	.23	.19	.29	.23	.19	.29	.23	.19	.27	.22	.18	.26	.22	.18	.17
9	.27	.21	.17	.27	.21	.17	.26	.20	.17	.25	.20	.16	.24	.19	.16	.14
10	.25	.19	.16	.24	.19	.16	.24	.19	.15	.22	.18	.14	.22	.18	.14	.13

MAXIMUM RECOMMENDED SPACING:

1.9 TIMES HEIGHT ABOVE WORKING PLANE

CANDELA AND LUMEN DATA ARE SHOWN FOR 1000 LUMENS AT THE SOURCE. EXPECTED PERFORMANCE VALUES FOR ANY OF THESE ITEMS MAY BE DETERMINED BY MULTIPLYING GIVEN VALUES BY AN APPROPRIATE FACTOR SUCH AS: LAMP LUMEN RATING X BALLAST FACTOR X MAINTENANCE FACTOR ÷ 1000 LUMENS

PER 1000 LAMP LUMENS

SHIELDING —	ANGLE BELOW HORIZONTAL
ARC TUBE	LAMP SHIELDED
GLASS ENVELOPE	AT ALL ANGLES

SUMMARY DATA			
ZONE	LUMENS	BARE LAMP	LUMINAIRE
90-180	29	3	4
0-90	686	69	96
0-60	533	53	75
0-30	137	14	19
0-180	716	71	100
TOTAL EFFICIENCY	71		

AVERAGE BRIGHTNESS	
VERT ANGLE	FT LAMBERTS*
85°	55
80°	66
75°	81
70°	103
[illegible]	129

*TO CONVERT TO CANDELAS/SQ INCH DIVIDE BY 452

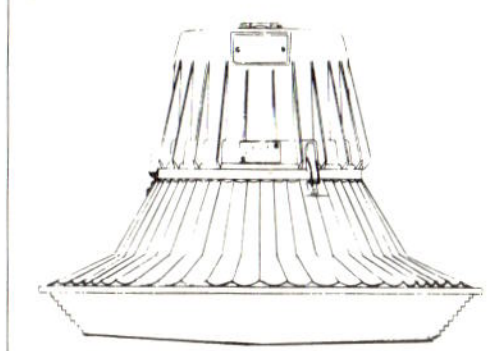

DISTRIBUTION DATA

VERTICAL ANGLE	CANDELA
180°	1
175°	1
170°	0
165°	0
160°	0
155°	0
150°	0
145°	0
140°	0
135°	0
130°	2
125°	5
120°	7
115°	8
110°	8
105°	8
100°	7
95°	9
90°	11
85°	18
80°	28
75°	42
70°	63
65°	92
60°	128
55°	160
50°	179
45°	184
40°	181
35°	177
30°	176
25	174
20°	166
15°	153
10	142
5	138
0°	137

ZONE	ZONAL LUMENS
180°-185°	
175°-180°	0
170°-175°	0
165°-170°	0
160°-165°	0
155°-°160	0
150°-155°	0
145°-150°	0
140°-145°	0
135°-140°	0
130°-135°	0
125°-130°	1
120°-125°	3
115°-120°	4
110°-115°	4
105°-110°	4
100°-105°	4
95°-100°	4
90°-95°	5
85°-90°	7
80°-85°	12
75°-80°	18
70°-75°	26
65°-70°	38
60°-65°	53
55°-60°	67
50°-55	75
45°-50	74
40°-45°	68
35°-40	59
30°-35	[illegible]
25°-30	44
20°-25°	36
15°-20	26
10°-15	17
5-10	10
0-5°	3

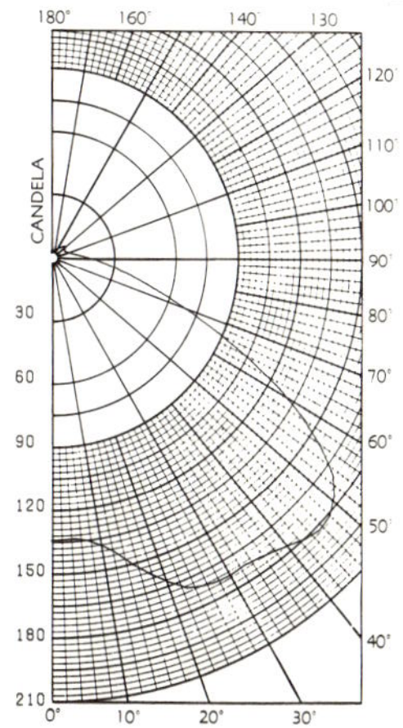

TESTED AT TWENTY-FIVE FEET WITH LAMP RATED AT 1000 LUMENS. IF DATA IS DESIRED FOR OTHER LAMP LUMENS MULTIPLY CANDELA AND LUMEN VALUES GIVEN BY RATIO — DIFFERENT LAMP RATING DIVIDED BY 1000 LUMENS

LUMINAIRE DESCRIPTION

LOWMOUNT INDUSTRIAL LUMINAIRE
FACETED SPECULAR REFLECTOR
STANDARD SOCKET POSITION
LAMP BASE DOWN

LAMP
400 WATT HIGH PRESSURE SODIUM
G.E. NO. LU400BD
ANSI CODE S51/BD

ANSI IES TYPE WIDE SPREAD

NUMBER

Fig. 4-1 Photometric data for a 400-watt high-pressure sodium luminaire designed to be used indoors at low mounting heights.Fig. 4-4

the performance of different luminaires and also ensures that the latest lamp lumen rating will be used in calculations.

The luminaire depicted on the graph has a symmetrical distribution; that is, the same candlepower is emitted in all directions at each vertical angle. Differences in distribution in the vertical plane are plotted on the graph. The luminaire candlepower in candelas is shown on the left of the graph. The numbers at the end of each radial line on the graph indicate vertical angles from the luminaire. The position straight down from the luminaire (the nadir) is 0 degrees, and the position straight up from the luminaire is 180 degrees.

When the distribution is asymmetrical, curves will be plotted for two or more vertical planes. These are 90 degrees apart for distributions which are elliptical, as shown in Fig. 4-2. For irregularly shaped distributions additional data may be obtained for planes at 45-degree intervals.

DISTRIBUTION DATA

The information given in the candela distribution graph is presented in tabular form as Distribution Data in Fig. 4-1. The two columns on the left give the candelas emitted from this luminaire in 5-degree vertical increments. Column 3 divides the vertical distribution from the luminaire into zones of 5 degrees each. The last column gives the zonal lumens in each of the 5-degree zones.

The distribution to each of the zones is calculated separately, since the

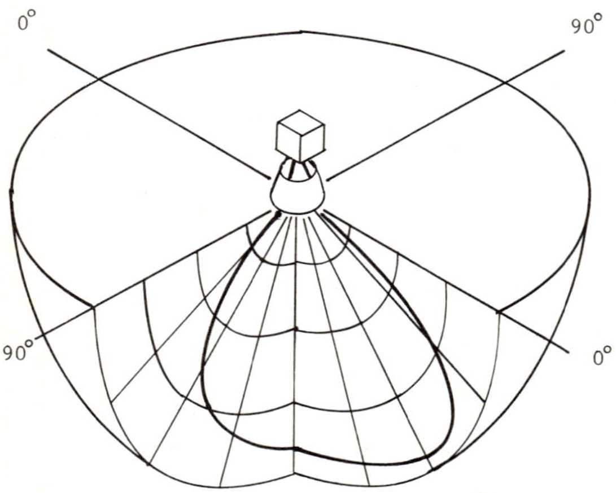

Fig. 4-2 When candela distribution is asymmetrical, curves can be plotted for two or more vertical planes. If distribution is elliptical, the curves are 90 degrees apart, as shown.

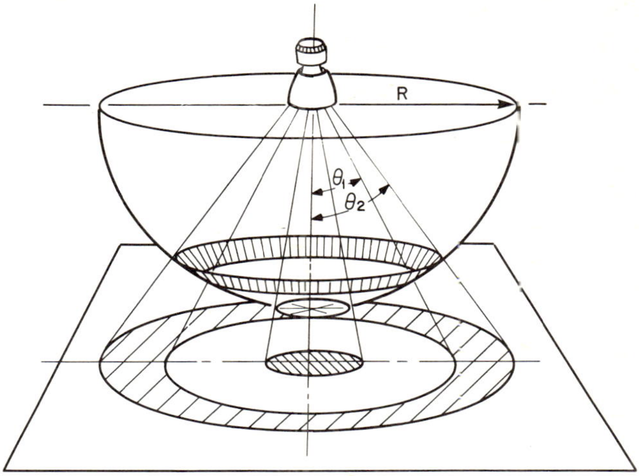

Fig. 4-3 Distribution data divides vertical distribution from a luminaire into 5-degree zones. This drawing shows how these zones look when projected on the work plane.

area of each zone gets larger as the vertical angle gets higher. Figure 4-3 shows how these circular zones look when projected on the work plane.

Sometimes zones smaller or larger than 5 degrees are used in calculations. If candela values change rapidly, smaller zones are necessary. Zones of 5 or 10 degrees are sufficiently accurate for industrial luminaires.

Calculations of zonal lumens are made on the assumption that the candela value in the center of each zone is the average of the candelas across a zone. Candela measurements are taken at the center of each zone; the lumens radiated from the luminaire are calculated by multiplying the candela value by a zonal constant.

The zonal constant which converts luminaire candelas to zonal lumens can be calculated by the formula

$$\text{Lumens} = \frac{\text{cd} \times \text{area}}{R^2}$$

where cd = candelas at the center of the zone
area = total area in each zone
R = test distance

The formula for the area of a spherical zone is

$$A = 2\,\pi\,R^2\,(\cos\theta_1 - \cos\theta_2)$$

where θ_1 = the vertical angle at the lower boundary of the circular zone
θ_2 = the vertical angle at the upper boundary of the circular zone

This is illustrated in Fig. 4-3.

The two formulas are combined to obtain the formula for a zonal constant. Since R^2 cancels out

$$\text{Lumens} = \text{cd } 2\,\pi\,(\cos\theta_1 - \cos\theta_2)$$

For the 5 to 10 degree zone, for instance, the zonal constant is 2π $(0.9961947 - 0.9848078) = 0.0715$.

This constant can be used to check how the value of 10 lumens was determined as the number of zonal lumens in the 5- to 10-degrees zone, as noted in Fig. 4-1. First, read the candelas at 5 and 10 degrees. At 5 degrees there are 138 candelas; at 10 degrees, 142. At 7.5 degrees, then, there are 140 candelas, which is the average of 138 and 142. Multiply this figure by the zonal constant of 0.0715, $140 \times 0.0715 = 10$, which is the number of lumens in the 5- to 10-degrees zone.

Distribution data is generally expressed per 1000 lamp lumens. The total number of lumens emitted from the luminaire is the sum of the lumens in each zone, multiplied by the lamp lumens divided by 1000.

Charts of zonal constants are published in the *IES Lighting Handbook.*

SUMMARY DATA

Figure 4-1 also gives Summary Data for this luminaire. This table summarizes the information given in the distribution data. The total number of lumens in various zones is given. The zone from 0 to 90 degrees is the direct component, which will be directed toward the work plane. The zone from 90 to 180 degrees is the indirect component, which will be radiated into the ceiling cavity. These two numbers combined give the total number of lumens from 0 to 180 degrees—in this case, 715.

The summary is given under two headings, Bare Lamp and Luminaire. The values for bare lamp are percentages of the 1000 lamp lumens used as the test base. The luminaire values are percentages of the total lumens radiated from the luminaire.

The ratio of 1000 lumens to the total radiated from the luminaire (in this case, 715) gives the efficiency of the luminaire, expressed as a percentage. This luminaire is 71.5 percent efficient.

The efficiency figure is not the same as the coefficient of utilization (CU). Other factors enter into the CU, such as the reflectance of the room and the distribution of the luminaire. The CU is a more significant figure than the efficiency, since the CU is used in calculating light levels. Sacrifices in luminaire efficiency are frequently made in order to improve the CU.

AVERAGE BRIGHTNESS

Photometric data usually includes a chart showing average brightness in the *glare zone,* which is considered to extend from the horizontal, or 90 degrees, down 35 degrees to the vertical angle of 65 degrees. Figure 4-1 includes the glare-zone chart for the sample luminaire. Photometric brightness is measured in footlamberts, which are units of luminance. Footlamberts are converted to candelas per square inch by dividing by 452, or to candelas per square meter by dividing by 0.292.

Luminaire brightness cannot be used to evaluate visual comfort, since it measures only the average brightness. Peak brightness is more of a problem where glare is concerned.

SHIELDING ANGLE

The shielding angle of a luminaire can be given in the form of a small chart, as shown in Fig. 4-1. The luminaire can be shielded at all angles or at a specific angle. The shielding angle is the angle between a horizontal line drawn across the bottom of the reflector and the line of sight at which the bare source first becomes visible. This is illustrated in Fig. 4-4. For high-intensity discharge (HID) or incandescent sources, shielding angles of 35 degrees or more are recommended because of the intensity of the source. The shielding angle is measured from the bottom of the arc tube for lamps having clear outer bulbs or from the bottom of the bulb when it is frosted or coated with a phosphor.

LUMINAIRE DESCRIPTION

One section of the photometric data gives the luminaire and lamp descriptions, as shown in Fig. 4-1. These include such information as the type of light source, part numbers, the lamp operating position, and the type of

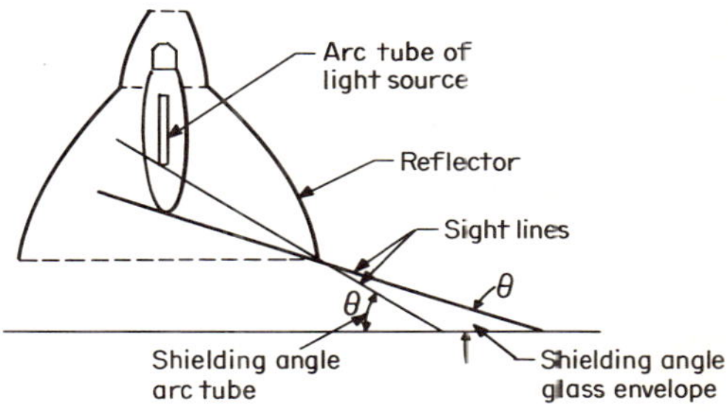

Fig. 4-4 The shielding angle of a luminaire is the angle at which the bare source first becomes visible. It is measured from the bottom of the arc tube for clear sources, or the bottom of the bulb if it is frosted or phosphor coated.

reflector design and finish. Also included is the position of the lamp, as determined by the socket setting. In the sample case, it is the standard socket position, since this luminaire comes in only one socket position. The lamp position tells where the light center of the lamp is in relation to the edge of the reflector.

In many luminaires, the lamp can be placed in several different positions. Each position changes the photometric distribution of the luminaire, as well as the shielding angle. Normally, the higher the lamp position in the reflector, the narrower the beam spread, and vice versa.

Lamp positions can be changed by placing spacers behind the lamp socket to move it in or out, or by a built-in adjustment method. Sometimes the spacer size is given in photometric data, such as "1 inch," "2 inch," etc.

The socket position is used mainly as a reference to identify the photometric data. Different HID lamp types and wattages have different light center lengths, and the socket position must therefore be changed to locate the different lamp types in the same relative position within the reflector. The socket position which produces a given photometric distribution is determined by the luminaire manufacturer at the time the reflector design is completed.

COEFFICIENTS OF UTILIZATION—ZONAL-CAVITY METHOD

Figure 4-1 gives the CU table for the same 400-watt industrial luminaire. The CU is the ratio of the lumens which fall on the work surface (usually a plane about 0.9 meters, or 3 feet above the floor) to the lumens generated by the lamp. The derivation of CU numbers and a more detailed explanation of CU tables are given later in this chapter.

The CU charts in photometric data are based on a luminaire spacing of 0.7 times the mounting height. This is a realistic spacing to use for high bay lighting, but it is often too close for systems installed at heights between 3 and 6 meters (10 and 20 feet). Any error in CU tables resulting from the use of a constant spacing to mounting height (S/MH) ratio is not significant.

MAXIMUM RECOMMENDED SPACING

Photometric data always include the manufacturer's recommendation of maximum spacing as a ratio to the mounting height above the work plane. This is probably the least understood of the values given in the data. Such ratings are frequently used to specify the beam pattern of a luminaire. They are used with the assumption that the quality of the lighting system remains constant regardless of the luminaire spacing so long as the S/MH rating is not exceeded. This is not always the case, especially if the task is specular, such as in the metalworking industry, and if visibility is influenced by both the luminaire brightness reflected in the task and the average illumination level.

To clear up some of the misunderstanding concerning S/MH ratings, the Design Practice Committee of the IES redefined the concept and published a new recommended practice. The redefined concept is called *luminaire spacing criterion* (SC) rather than *maximum S/MH ratio.*

The report of the committee was published in August 1977 in *Lighting Design and Application,* the magazine of the IES. The following explanation of luminaire SC is taken from that report with the permission of the IES. Limitations noted in the section after the explanation are the authors' own thoughts.

Explanation of Spacing Criterion

Luminaire SC is a classification technique for interior luminaires relating to the photometric distribution provided by the direct component (0 to 90 degrees) from the luminaires. The technique is a quick way of giving the designer one necessary piece of information. Spacing criterion should not be used as a specification for the actual ratio of the spacing to the mounting height of the luminaires. In this analysis, mounting height is considered to be measured from above the work plane.

FORMULATION

The concept of luminaire SC is based on the assumption that all of the principal horizontal illumination on the work plane directly under a particular luminaire is provided by that luminaire. Each luminaire is expected to contribute enough light so that point Q in Fig. 4-5 midway between luminaires, will be illuminated to the same level as point P, directly beneath one of the luminaires. This assumes that each luminaire will contribute half as much illumination to point Q as it does to point P.

The point midway between luminaires or in the middle of a square formed by four luminaires can also be a point of low illumination. However, if the procedure for luminaire SC is followed, all of these points will have the same illumination as the point directly under the luminaire. In Fig. 4-5 point R is the midpoint of four luminaires and point P is directly beneath a luminaire. In order for the levels of illumination at R and P to be equal, each of the four luminaires must contribute one-fourth as much illumination to point R as it does to point P.

The goal, then, is to provide points R, P, and Q with uniform illumination. By using the graph in Fig. 4-6, it is possible to determine the spacing at which any luminaire meets these specifications. The only assumption for the following computations is that the inverse-square law is valid.

PROCEDURE

The following procedure is quoted exactly, with permission, from the IES recommended practice *Classification of Interior Luminaires by Distribution: Luminaire Spacing Criterion,* LM-53.

A. For luminaires whose intensity distribution is nominally symmetric about the nadir:

1. Plot the relative intensity of the luminaire on the chart [in Fig. 4-6].

2. Locate the point of one-half the intensity at 0 degrees on the ordinate, and draw a line through that point, parallel to the diagonal lines. If the intensity varies significantly in the vicinity of 0 degrees, use an average of the intensity over the 0 to 5 degrees polar angle.

3. Read scale A above the intersection of this line with the intensity curve.

4. Repeat step 2, using the point of one-quarter the intensity at 0 degrees.

5. Read scale B above the intersection of this line with the intensity curve.

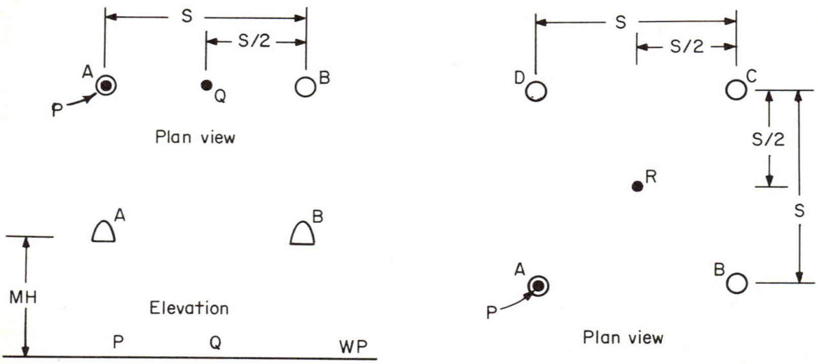

Fig. 4-5 Point Q, midway between two luminaires, and point R, the midpoint between four luminaires, could be points of low illumination. If SC recommendations are followed, points Q, R, and P will have equal illumination. (These drawings are reproduced with permission from the IES publication *Classification of Interior Luminaires by Distribution: Luminaire Spacing Criterion,* LM-53.)

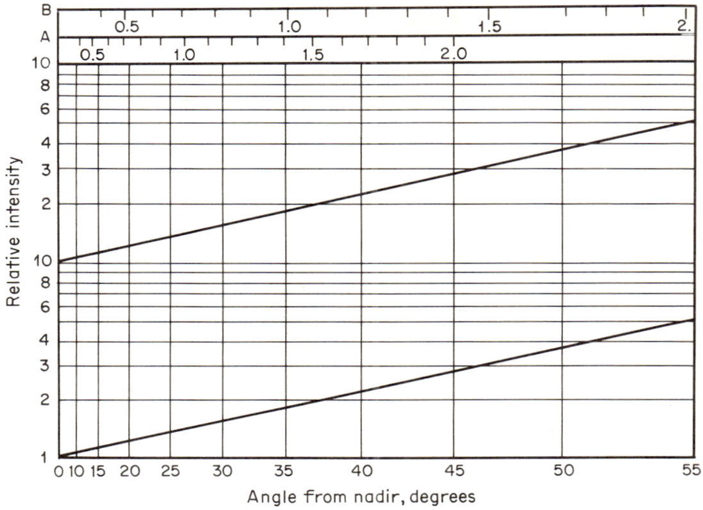

Fig. 4-6 Graph used to determine SC. (Reprinted with permission from the IES publication *Classification of Interior Luminaires by Distribution: Luminaire Spacing Criterion,* LM-53.)

6. The lower of the values found in steps 3 and 5 shall be taken as the luminaire spacing criterion. Round off the value to the nearest 0.1.

B. For luminaires with significantly asymmetric intensity distributions about the nadir:

Independently evaluate the longitudinal and transverse (parallel and perpendicular, 0 degrees and 90 degrees, etc.) intensity distributions. Apply steps A.1, A.2, and A.3 for each of the intensity curves. Round off to the nearest 0.1.

INTERPRETATION

The value from scale A corresponds to the criterion of point Q, the point midway between two luminaires. The value from scale B corresponds to the criterion of point R, the midpoint between four luminaires. For a luminaire with symmetrical distribution, the level of intensity at points Q, R, and P should be equal.

For an asymmetric luminaire, independent testing at point Q for each orientation will be adequate.

The luminaire SC suggests a maximum spacing at which the horizontal illumination will be reasonably uniform. This is a nominal value, and in most cases a closer S/MH ratio will provide better results.

The recommended practice suggests that 1.5 be the maximum value for luminaire SC, because of additional factors, such as overlap between luminaires, vertical illumination, shadowing, illumination distribution above the work plane, etc.

Limitations of Spacing Criterion

Because of the number of different beam shapes available and the varying conditions which exist within an industrial facility, any single-number SC or suggested spacing will by itself fall short of providing a good indication of how well a luminaire will perform. The recommended practice cautions that the SC number should not be used as the S/MH ratio and that generally an S/MH ratio less than the SC produces better results. Other factors that affect the total performance of the lighting system also need to be evaluated.

Nevertheless, such single-number classifications are often written into specifications without any clarification. This is done in some cases to identify the beam spread of the luminaire but may erroneously be used in the design of the system.

In addition to the factors named above, there is a major limitation in the use of the SC number as a spacing guide in an industrial plant. By definition in the recommended practice, each luminaire provides all of the principal light directly beneath, and half the light midway between, luminaires. Such an installation would produce good results if there were no obstructions or shadows to limit the illumination from any one luminaire.

But this description does not fit the average work environment in an industrial facility. As a result, spacing luminaires at the distance given by their SC would produce marginal results. A better approach is to design a system in such a way that every point on the work plane is illuminated by at least two luminaires. In this way, if light from any one luminaire is blocked by equipment or persons working in the area, there will still be some light on the work plane. This is an especially critical factor for persons working with metal surfaces, since reflected light from the luminaire must be visible in the surface in order to produce surface brightness.

CALCULATING RELATIVE CONTRIBUTIONS

In order for each part of the work plane to be illuminated by more than one luminaire, the beam spread must be sufficiently wide so that some light reaches the area under each adjacent luminaire. For purposes of discussion, consider the level of illumination to be constant, since these comments concern the *relative* contribution of each luminaire to the illumination on a particular spot.

Figure 4-7 illustrates a simple layout. It consists of nine luminaires, with one in the center surrounded by eight others in a square pattern. For best results, luminaires A and B should contribute some luminous intensity to the area under the center luminaire C. In each of the following examples, assume that luminaires of the same type and wattage are used in the A, B, and C locations.

Luminous intensity must be directed at certain vertical angles in order

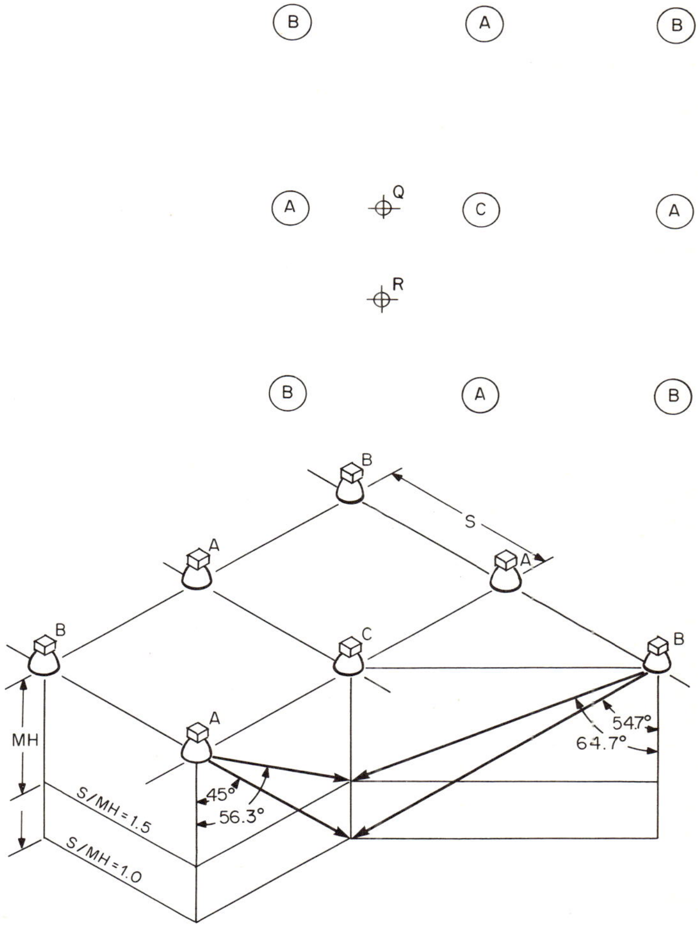

Fig. 4-7 Layout of nine luminaires in a square pattern. The elevation gives the angles at which candelas must be emitted in order to reach the area under adjacent luminaires at S/MH ratios of 1.5:1 and 1.0:1.

to provide illumination under adjacent luminaires. The particular vertical angle varies with the S/MH ratio, as indicated in the elevation shown in Fig. 4-7.

For example, assume that the luminaires in line with one another are spaced at a 1:1 S/MH ratio. In this situation, the A luminaires must provide some luminous intensity at 45 degrees in order to reach the area under luminaire C. The B luminaires are on the diagonal from C and 40 percent further away than the A luminaires; the B luminaires must therefore provide some luminous intensity at 54.7 degrees.

The same graph which is used to calculate SC can be used to determine the contributions which the A, B, and C luminaires make to the area under C. Figure 4-8 is an expanded version of the graph in Fig. 4-6. When selecting SC, the A and B charts at the top of the graph are used. In this discussion the SC is already known and the purpose is to determine relative luminous intensity.

As noted in the recommended practice, the graph is based on the inverse-square law:

$$E = \frac{\text{cd} \cos^3 \theta}{\text{MH}^2}$$

Luminous intensity directed straight down (the nadir, or 0 degrees) will provide a higher level of illumination than the same intensity directed out at some vertical angle. The graph can be used to determine the number of candelas at each vertical angle that would produce the same level of illumination as a given number of candelas at the nadir; that is, the two differing candela values would produce the same light level, one at a given vertical angle and the other at the nadir.

The diagonal lines represent constant illumination away from the luminaire. If the candela distribution of a luminaire exactly matched one of the diagonal lines, then the light level produced by that luminaire would be uniform at each point on the work plane.

Plotted on the expanded graph are the distributions for a family of three luminaires having spacing criterions of 1.0, 1.5, and 2.0. Any convenient spot on the left ordinate can be used as a starting point. In this case, each number stands for 1000 candelas. Distribution data for each luminaire was used to find the number of candelas at each vertical angle.

Luminaire with 1.0 SC If the luminaire with the 1.0 SC were placed at an S/MH ratio of 1:1, the A luminaires in Fig. 4-7 would have to contribute light at 45 degrees in order to reach the area under C.

Figure 4-8 can be used to find the relative level of illumination produced by this luminaire at 45 degrees. To read the graph, find the 45-

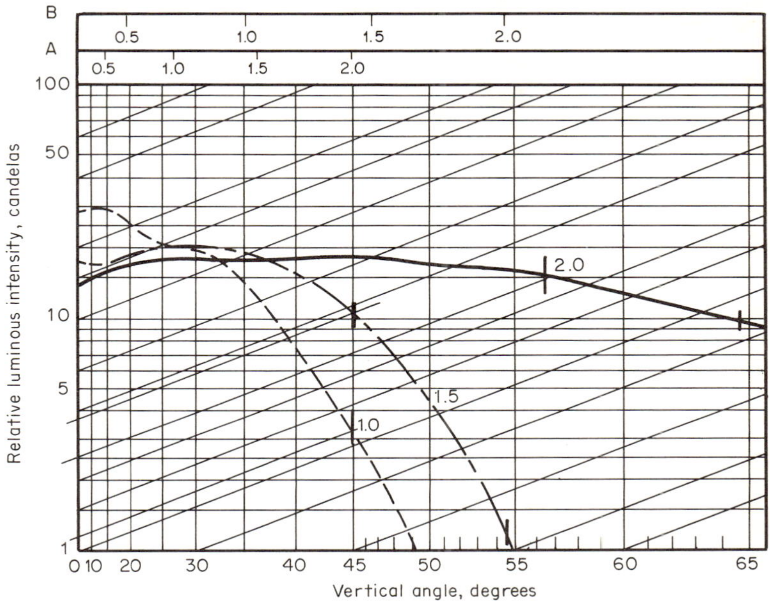

Fig. 4-8 This expanded graph can be used to compare relative luminous intensity in candelas for luminaires with differing candela distributions and SCs.

degree point on the lower ordinate. Follow the line up to the point where it crosses the candela distribution plot of the 1.0-SC luminaire. There was no diagonal line at this point, and so one was drawn parallel to the other lines. Follow this line back to the 0-degree line and read the number of nadir candelas: 1200 candelas emitted from a luminaire directly overhead will produce the same level of illumination as 3300 candelas directed at 45 degrees.

The four A luminaires would contribute 1200 candelas each to the area under C. The C luminaire would contribute 28,600 candelas at the nadir, as indicated on the graph. The B luminaires would have to contribute light at an angle of 54.7 degrees in order to reach the area under C, and the 1.0-SC luminaire emits very little light at that angle. The B luminaires, then, would make a negligible contribution.

The total contribution to the area under C would thus be equivalent to $(4 \times 1200) + 28{,}600 = 33{,}400$ candelas. The A luminaires would contribute $4800/33{,}400 = 0.14$, or 14 percent, of the light level.

If this system were installed and the luminaire overhead (C) were somehow blocked by machinery or a worker's shadow, the level of illumination would drop by 86 percent. This calculation takes into account only the direct contribution of each luminaire, but in many factories with low surrounding-surface reflectance, the direct contribution is the principal source of illumination.

The luminaire with the 1.0 SC, spaced at 1:1, would not be a satisfactory choice for an industrial application, unless the area were open, with little chance of light output being obstructed.

Luminaire with 1.5 SC Figure 4-8 shows that the luminaire with the 1.5 SC should not be spaced at a 1.5 S/MH ratio in a crowded industrial facility, where there are difficult tasks and light may be obstructed. If 1.5:1 were the S/MH ratio, light from the A luminaires in Fig. 4-7 would need to be directed at an angle of 56.3 degrees in order to reach the area under C. There is virtually no contribution from this luminaire at 56.3 degrees. The B luminaires would need to project light at an angle of 64.7 degrees, and this luminaire directs no light at all at so high an angle. This means that luminaire C would provide all of the principal illumination beneath it.

If the luminaire with the 1.5 SC were installed at an S/MH ratio of 1:1, however, these angles would be reduced, as shown in Fig. 4-7. The contribution from the A luminaires would need to be directed at 45 degrees; at that angle, this luminaire contributes the equivalent of 3800 nadir candelas. This value is read in Fig. 4-8 by finding the point where the 1.5-SC plot crosses the 45-degree line and then following the diagonal line back to the 0-degree line.

The 1.5-SC luminaire provides 17,500 candelas at the nadir. The total equivalent contribution to the area under C would thus be (4 × 3800) + 17,500 = 32,700 candelas. The relative contribution of the A luminaires would be 15,200/32,700 = 0.46, or 46 percent.

Since the B luminaires would make a small contribution to the illumination level, the contribution of the eight luminaires around C to the area under C would be approximately 50 percent. The luminaire overhead would provide the other 50 percent. If light from any one luminaire were blocked or a lamp failed, each worker would still have enough illumination to continue performing the necessary tasks.

Luminaire with 2.0 SC Similarly, the luminaire with the 2.0 SC would produce better results if spaced at a 1.5:1 S/MH ratio than if spaced at an S/MH ratio of 2:1. Reading from Fig. 4-8, the 2.0-SC luminaire provides 13,700 candelas at the nadir. At 56.3 degrees—the angle needed at 1.5 S/MH—the A luminaires would provide the equivalent of 2500 nadir candelas. The relative contribution of the four A luminaires would be

$$\frac{4 \times 2500}{(4 \times 2500) + 13{,}700} = 0.42, \text{ or 42 percent}$$

Since the B luminaire would make almost no contribution, the C contribution would be 58 percent. This would produce satisfactory results.

Contribution to Q and R Point Q midway between two luminaires and point R at the midpoint of four luminaires (Fig. 4-7) would receive adequate illumination if this method of determining spacing were followed. Since the area under each luminaire would be lighted from more than one direction, it follows that areas between luminaires would also be lighted from more than one source.

CONCLUSIONS

The major limitation of the SC concept is the recommendation that the point directly beneath a luminaire receive almost all of its illumination from the luminaire directly overhead. As indicated in the examples, this limitation can be overcome if luminaire spacing is 0.3 to 0.5 smaller than the rated SC of the luminaire. In this way, the contribution of the luminaire directly overhead would be about 50 percent and the contribution from adjacent luminaires would be about 50 percent.

For instance, if a 1:1 S/MH ratio is desired, then a luminaire with an SC of 1.3 to 1.5 can be used successfully. For a 1.5-S/MH installation, the SC of the luminaire chosen should be 1.8 to 2.0.

Luminaire spacing above 1.5:1 is not recommended in a manufacturing facility where individual work stations exist. This is especially true in the metalworking industry.

At a 1.5:1 S/MH ratio there will be a luminaire directly or nearly overhead for each work station. The worker will thus not be too conscious of where the light is coming from. In this case, there will be sufficient overlap between luminaires.

If luminaires are spaced at an S/MH ratio of 1.8:1 or 2.0:1, however, a luminaire with an SC over 2.0 will be needed. At a 2.0:1 S/MH ratio, luminous intensity from adjacent luminaires has to occur at 63 to 70 degrees. This could result in high luminaire brightness and a potentially glaring installation. Luminaires have to get brighter as S/MH is increased, because the lamp image is lower on the reflector when viewed from any given viewing angle.

The brightness problem can be overcome by using a diffuser to shield the light source. This is often done when luminaires are designed with very wide beam spreads, to be used at low mounting heights.

There is, however, a drop in the CU as beam spreads are made wider. The most satisfactory installation will be one which uses the luminaire with the best compromise between uniformity, CU, and brightness.

In many locations, luminaires can be spaced at an S/MH ratio equal to their SCs with no problem. In areas where storage or material handling are the only tasks, spacing wider than 1.5:1 could be acceptable.

Most of the problems with luminaire beam spreads and SC occur at low mounting heights, less than 7.6 meters (25 feet). Above 7.6 meters, in what is commonly classified as high bay lighting, luminaire spacing is usually under the value given for the SC. Here, too, using a luminaire with an SC slightly above the actual S/MH of the installation will give better results.

Derivation of the Coefficient of Utilization—Zonal-Cavity Method

The CU is so integral a part of lighting calculations that an understanding of the derivation of CU tables is helpful. The zonal-cavity method of determining the CU is based on the concept that the area to be lighted consists of a series of spaces or cavities which have effective reflectances with respect to one another and to the work plane. The direct contribution of the luminaire is combined with the light reflected from the walls and from the floor and ceiling cavities to determine the total illumination on the work plane. The CU compares this illumination on the work plane with the initial lamp lumen output.

In order to read a manufacturer's CU tables the designer needs to gather information about the particular area to be lighted. The room cavity ratio, wall reflectance, and effective ceiling and floor cavity reflectances must be known as well as the tentative lamp and luminaire choice.

ROOM CAVITIES

To determine the CU, the designer divides the space to be lighted into three cavities, as shown in Fig. 4-9. The ceiling cavity is the space between

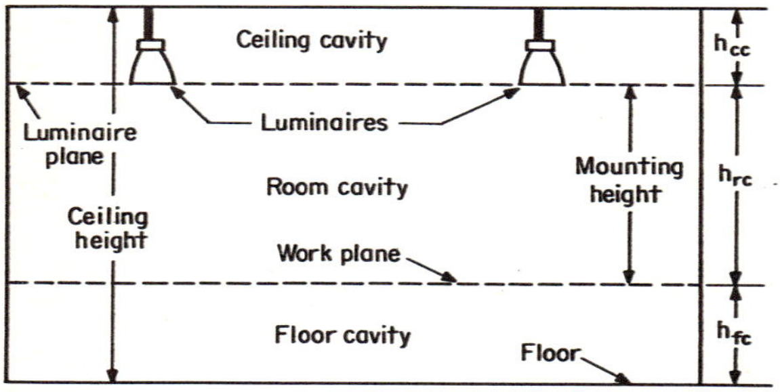

Fig. 4-9 To derive CU tables, the designer divides the space to be lighted into three cavities. These zonal cavities have effective reflectances with respect to each other and to the work plane.

Fig. 4-10 This area was divided for design purposes. Note that two different types of luminaires have been used.

the ceiling and the luminaire plane (the bottom of the luminaires). The room cavity is the space between the luminaire plane and the work plane. The floor cavity is the space between the work plane and the floor.

If the ceiling, floor, or room area is broken up with beams, ducts, machinery, or other obstructions, then the area to be lighted should be broken down into smaller sections for design purposes. Fig. 4-10 is a photograph of an industrial facility which was divided into smaller areas for the selection of appropriate lighting equipment. Note that two different types of luminaires have been used in the two main areas visible in the photograph.

CAVITY RATIOS

When the geometric proportions of the ceiling, room, and floor cavities are calculated, the results are numerical relationships called *cavity ratios.* The three cavity ratios can be determined with this formula:

$$\text{Cavity ratio} = \frac{5h\ (\text{room length} + \text{room width})}{\text{room length} \times \text{room width}}$$

where h = height of the cavity being measured. As shown in Fig. 4-9, h_{cc} is the height of the ceiling cavity; h_{rc} is the height of the room cavity; and

h_{fc} is the height of the floor cavity. The three ratios derived with the formula are the *ceiling cavity ratio* (CCR), the *room cavity ratio* (RCR), and the *floor cavity ratio* (FCR).

If the luminaires are flush with the ceiling or recessed, the CCR is 0. If the floor is the work plane, the FCR is 0.

If the area to be lighted is square, the RCR can be found with the formula

$$\text{RCR} = 10 \times \frac{\text{height}}{\text{width}}$$

If the area is irregularly shaped, cavity ratios can more easily be determined with this formula:

$$\text{Cavity ratio} = 2.5 \times \frac{\text{total cavity wall area}}{\text{area of cavity base}}$$

The wall area for each wall is calculated by multiplying the length times the height. Areas of all four walls are added to yield the total cavity wall area used in the formula.

The three cavity ratios are interrelated, so that

$$\text{CCR} = \text{RCR} \times \frac{h^{cc}}{h^{rc}} \qquad \text{and} \qquad \text{FCR} = \text{RCR} \times \frac{h^{fc}}{h^{rc}}$$

Once the RCR is known, the other ratios can be computed easily.

Figures for cavity ratios can be obtained by using these formulas or by referring to charts published in the *IES Lighting Handbook.* They can also be determined by using the graph given with the sample problem in Chap. 5.

REFLECTANCES

In an indoor application, all lumens radiated from a luminaire must eventually be absorbed, since the area is closed; but as more incident light is reflected in the room, more radiant energy occurs in the space at any one time. In general, the higher the reflectances, the higher the utilization of the light.

It is necessary, then, to determine the reflectances of ceilings, walls, and floors. Expected maintained values should be used for ceiling, wall, and floor reflectances if maintained levels of illumination are to be calculated. Initial figures can be used if initial illumination is to be determined.

Reflectances can be estimated or measured. If the wall of the room cavity in question is a large piece of machinery or a stack, some reflectance value must be estimated or measured for that surface in order for calculations to be accurate.

Fig. 4-11 In some cases it is almost impossible to estimate reflectances. In the area shown, there are so many crossbars and trusses that it would be difficult to determine accurate reflectance values.

Sometimes, determining reflectances is virtually impossible, as in the facility in Fig. 4-11. If reflectances are unknown, a 30 percent wall, 30 percent ceiling, and 20 percent floor reflectance can be used. When the building is new, these values will be higher, but they will deteriorate rapidly in most industrial situations. These estimates are very low because wall surfaces in factories are not usually well maintained.

Reflectances in office areas are generally higher than those in manufacturing or storage facilities. Information in Chap. 8 should be used for office lighting design.

EFFECTIVE REFLECTANCES

Reflectance values must be adjusted to yield effective reflectances. The CCR and FCR values are used to adjust downward the reflectance of the floor or ceiling if the cavity is deep. The depth of the ceiling cavity influences how much of the light reflected off the ceiling gets back to the work plane. In a deeper cavity less light is reflected back. This has the same effect as a reduction in the reflectance of the ceiling surface. The same holds true of the floor cavity.

Wall reflectance is involved in the effective reflectances of both ceiling and floor because a portion of the light reflected in the ceiling or floor cavity is bounced off the walls. Tables such as the one given in Chap. 5 can be used to combine wall reflectances with the CCR or FCR and the ceiling or floor reflectance to find the effective reflectance of either the ceiling or floor cavity.

COEFFICIENT OF UTILIZATION CHARTS

The effective ceiling cavity reflectance and the percentage wall reflectance are combined with the RCR in CU charts, such as the one given earlier. These charts assume a 20 percent effective floor cavity reflectance. If the effective reflectance is not 20 percent, refer to tables printed in the *IES Lighting Handbook.* These give multiplying factors that yield effective reflectances for floors with other than 20 percent reflectances.

The zonal-cavity method assumes that room surfaces are uniform and diffuse. In industrial plants, the reflectance of room surfaces is rarely over 50 percent and can be much lower if dirt has accumulated on the surfaces; room surface reflectance will therefore have only a limited effect. Even if room surfaces are not uniform and diffuse, CU charts can still be used with reasonable accuracy.

In the next chapter a method is given for determining how much light is reflected from a wall and contributes to any point on the work plane. This procedure can be followed if more accuracy is desired.

Source Selection

Photometric data can aid in the selection of sources and luminaires. The designer must make a tentative selection before proceeding with lighting design calculations.

The HPS lamp is the most efficient industrial lighting source, with efficacy of up to 140 lumens per watt. It can be used in any indoor or outdoor industrial application, except perhaps for very close color-matching. There are a number of sizes, and the illumination level can therefore be matched with the number of luminaires needed to provide good uniformity. If the newer, electronic regulator ballast is used with these sources, lamp lumen output remains virtually constant, so that if luminaires and room surfaces are cleaned regularly, light levels will also remain close to 100 percent of initial values.

The next chapter includes a quick way of choosing lamp wattage according to initial rated lumen output. This choice is intended as a first estimate and must be verified through the design process.

Luminaire Selection

A tentative luminaire choice should also be made before the lighting system calculations are begun. Luminaire selection is based on a number of factors, including the ambient environment, mounting height, beam spread, shielding angles and reflector shape, and standby or emergency needs.

AMBIENT ENVIRONMENT

The environment in which a luminaire will be placed is a major factor in luminaire selection. Hazardous locations where there are potentially explosive vapors or easily ignitable fibers or flyings require specially designed equipment.

Wet or corrosive environments also require special luminaires. Gasketing, special paints, and other features of luminaires designed for this kind of service are described in Chap. 3.

Temperature is another factor; some ballasts will fail or not operate at very high or very low temperatures.

If the luminaire will be subject to vibration or mechanical shock, this too must be considered. Whether the environment is basically clean or dirty is another factor. Sealed and filtered luminaires should always be used in dirty ambient environments.

MOUNTING HEIGHT

Some luminaires are designed for use primarily at low mounting heights, that is, below 7.6 meters (25 feet). Other luminaires are intended for use primarily above 7 meters. These can be installed in high bay areas. Many of these luminaires can be used at lower mounting heights as well.

BEAM SPREAD

There are a number of different beam spreads available with any one HID luminaire. In most cases, a particular type luminaire is part of a family of HID luminaires having the same general characteristics.

Beam-spread variations within a family of luminaires can be accomplished by changes in lamp position within the reflector. Usually two or more reflector sizes are needed for the complete range of low, medium, and high bay applications. Generally, moving the lamp lower in the luminaire will result in a wider beam spread.

Beam Spread and Spacing Criterion Wider beam spreads mean that luminaires can be spaced further apart. Figure 4-12 is a graph of one family of HPS luminaires. Each beam spread is marked with its recommended SC. The bottom ordinate indicates the vertical angle from the luminaire, as well as the S/MH ratio.

When beam spread is increased, as indicated by a higher SC, nadir candlepower is reduced; that is, the number of candelas directed straight down at 0 degrees is less. In Fig. 4-12, the 1.0-SC luminaire has the narrowest beam spread. This luminaire delivers more than twice as much illumination to the point directly beneath the luminaire as the 1.9-SC luminaire, which has the widest beam spread. These are relative values and will vary with different lamp types and reflector designs.

The change at the nadir is much greater than the change beyond 35 degrees. For this reason improved uniformity in any situation results

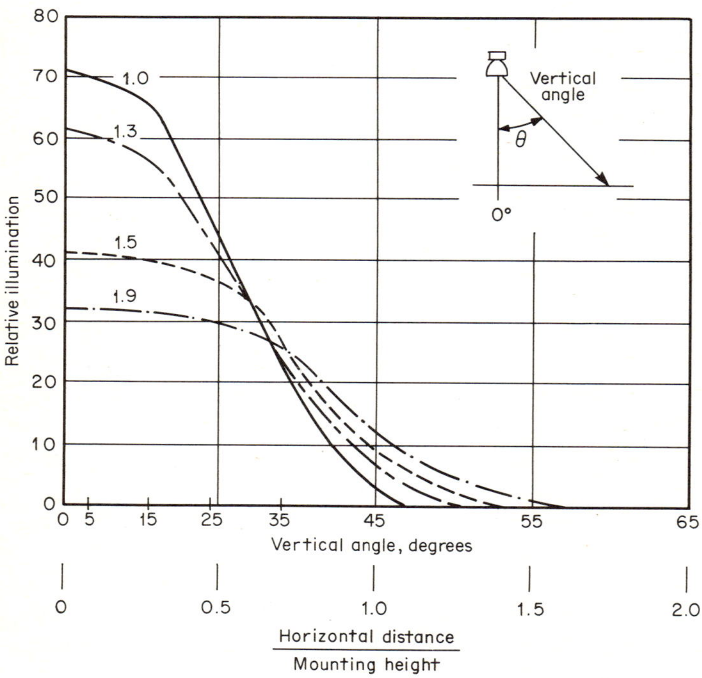

Fig. 4-12 Graph of relative illumination for a family of HPS luminaires. Improved uniformity in any situation results more from a reduction of the lighting level directly beneath a luminaire than it does from increasing the level between luminaires.

more from a reduction of the lighting level directly beneath the luminaire than from an increase in the level between luminaires.

The next chapter includes a quick method of estimating maximum nadir candlepower for a given situation. This can be used to make a tentative luminaire choice. The choice must then be verified through the design procedure.

It is easy to provide illumination immediately under a luminaire. At the nadir the entire surface of the reflector is available to redirect the lamp's output. The distance between the luminaire and the work surface is also the shortest at this point. As the vertical angle increases, the projected reflector area decreases in proportion to the cosine of the vertical angle. At 90 degrees, the reflector is completely cut off. Thus, the provision of light at high angles is limited by the projected area of the reflector.

The discussion of SC limitations showed that light must be projected at 45- to 54-degree angles in order for more than one luminaire to contribute illumination to each point on the work plane. This is therefore a major consideration in the choice of luminaires.

Beam Spread and Utilization The choice of luminaires with higher SCs and wider beam spreads must be balanced against the CU. As the RCR gets larger, wide-beam luminaires lose utilization faster than narrow-beam luminaires.

Figure 4-13 shows the effect of the RCR on the CU for three beam spreads, with SCs of 1.0, 1.3, and 1.5. Room reflectances are assumed to be 30 percent for the ceiling cavity, 30 percent for the walls, and 20 percent for the floor cavity.

The 1.0- and 1.3-SC curves overlap between RCRs of 5 and 6. The 1.0- and 1.5-SC curves overlap at an RCR of 4.

If the curves were plotted on the basis of equal lumens in the downward direction, the narrower distribution would show a higher utilization in all cases. In practice this is not always the case, because the luminaire becomes more efficient as the beam spread is made wider. This is caused by the lamp being brought closer to the lower edge of the reflector. If two luminaires have the same utilization, the one with the wider beam spread should be chosen.

Choosing the proper beam spread is usually a compromise between the need for wider beam spreads, improved uniformity, and the CU. The best choice can be made by comparing the performance of a family of different beam spreads and choosing the one with the highest utilization consistent with the need for sufficient overlap between adjacent luminaires.

Beam Spread and Levels of Illumination Each luminaire produces a different pattern of illumination on the work plane. The next chapter,

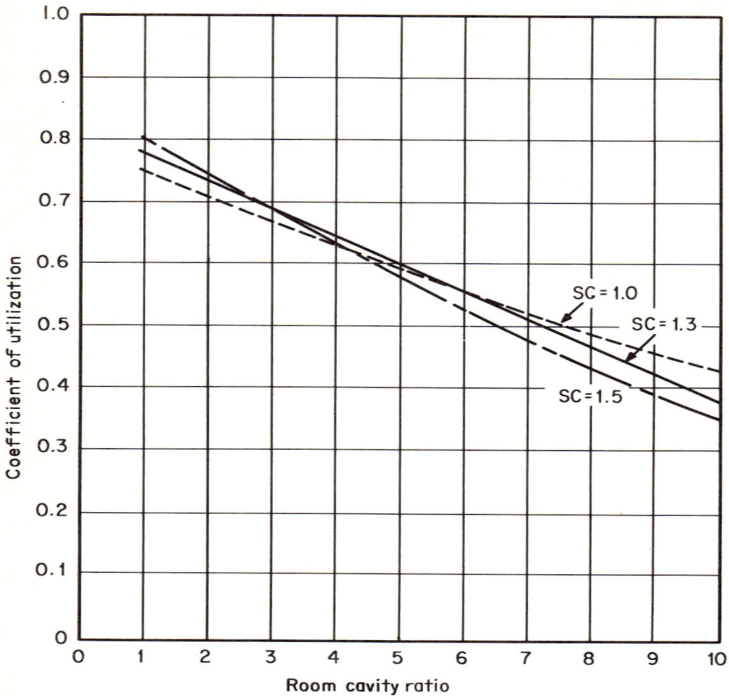

Fig. 4-13 Relation of the CU to the RCR. The open-reflector room reflectances are: ceiling, 30 percent; walls, 30 percent; floor, 20 percent.

Chap. 5, includes a graph that can be used to determine the level of illumination at any point on the horizontal plane with a particular luminaire and mounting height.

SHIELDING ANGLE AND REFLECTOR SHAPE

Lamps should be shielded so that they are not visible at normal viewing angles, particularly at low mounting heights. This usually requires a shielding angle of at least 35 degrees.

The wider the beam spread, the brighter a luminaire reflector will appear. In practice it is desirable for the flashed region of the reflector to be visible first at the same angle at which the lamp can first be seen (the shielding angle). (The flashed area is the area where the lamp brightness can be seen reflected in the reflector's surface.) As the beam spread is increased, this flashed portion must occur lower on the reflector surface. With smooth reflectors the lamp image on the reflector surface will be almost as bright as the lamp itself, since it occurs in a narrow band directly behind the lamp.

The peak brightness can be reduced by etching the reflector surface or by pressing a pattern into the reflector that breaks up the lamp image (see Chap. 3). Increasing the reflector size will not significantly ease the brightness problem unless some means are used to spread the brightness over the larger reflector surface.

Brightness control can also be achieved by putting a diffuser between the lamp and the room. The size of the diffuser must be such that its average brightness is not offensive. Prismatic control can also be introduced into the diffuser for added brightness control.

EMERGENCY OR STANDBY SITUATIONS

High-intensity discharge sources have a slow warm-up time, and so some form of auxiliary lighting is usually required in conjunction with these systems. One method of providing this lighting is to install small quartz incandescent lamps in about 10 percent of the HID luminaires. These lamps light immediately when power is restored after an interruption and remain lighted until the HID sources have reached approximately 60 percent of their full light output.

Another approach is to use separately mounted fluorescent or incandescent sources as standby lighting.

Emergency lighting is needed to provide enough illumination for workers to leave a building in the event of a power failure. This is an entirely separate lighting system, as explained in Chap. 14.

Lighting Calculation Methods

Once tentative lamp and luminaire choices have been made, the designer can proceed with lighting calculations. There are two methods commonly used to calculate illumination in enclosed indoor spaces. One is the lumen method, which incorporates the zonal-cavity approach to obtaining the CU, and the other is a point-by-point approach utilizing the inverse-square and cosine laws. The lumen method gives an average horizontal illumination level in the space, and the point-by-point method determines the level of illumination at a particular point on either a horizontal or a vertical surface.

When lighting an interior space, the designer considers the contribution of interreflected light along with the direct contribution of the luminaire. Interreflectance is included in the CU tables used in the lumen method, but it is not a part of most point-by-point calculation methods. There is a way of determining the point-by-point contribution of the reflectance from a wall, and this is given in the next chapter.

Computer programs have been evolved which successfully combine averaging and point-by-point techniques for office lighting design. Computer programs that calculate point-by-point illumination levels in industrial plants are also available. There will eventually be programs which take into account the interreflected contribution as well as shadows caused by obstructions. This could significantly improve the accuracy of predicting the illumination at any point on the work plane.

LUMEN METHOD

The lumen method is a way of calculating the expected level of illumination on a horizontal plane from a particular lamp-luminaire combination. It is based on the definition of illumination as luminous flux per unit area, where luminous flux is expressed in lumens.

This chapter will explain this approach to lighting design. A sample problem is solved with the lumen method in the following chapter, Chap. 5.

The formula for calculating expected light levels by the lumen method is

$$E = \frac{N \times \text{LL} \times \text{CU} \times \text{LLF}}{\text{area}}$$

where E = illumination
N = number of luminaires
LL = initial rated lamp lumens
CU = coefficient of utilization
LLF = light-loss factor
area = size of the area to be lighted

Illumination Level (E) Recommended illumination levels can be obtained from the charts in Chap. 1. These are minimum maintained levels; initial levels must be higher.

Number of Luminaires (N) The formula can be rewritten to solve for the number of luminaires to be used, rather than for E. It would then read

$$N = \frac{E \times \text{area}}{\text{LL} \times \text{CU} \times \text{LLF}}$$

Lamp Lumens (LL) This is the initial rated lamp lumens for a particular lamp. This figure is available from manufacturers.

Coefficient of Utilization (CU) The CU is the ratio of lumens reaching the work plane to the total lumens generated by the lamp. The CU takes into account interreflectances of light in the room, the efficiency and distribution of the luminaire, its mounting height, and room proportions. Derivation of the CU was explained earlier in this chapter.

Light-Loss Factor (LLF) Light-loss factors are those factors which contribute over a period of time to a decline in the light output of a given lamp-luminaire-ballast combination. The light-loss factor is usually considered to be lamp lumen depreciation (LLD) times luminaire dirt depreciation (LDD), but some of the other components are voltage to the luminaire, ballast factor, luminaire reflector depreciation, room-surface dirt depreciation, and lamp outages tolerated.

Ballast factor is the percentage of lamp lumens produced by a ballast. Lumen output is measured in laboratories under a specific voltage and current which produce the specified wattage. High-intensity discharge ballasts are designed to have a 100 percent ballast factor, if the lamp-luminaire-ballast combination operates at the nominal rating recommended by the manufacturer.

Lamp lumen depreciation is obtained from manufacturer's data. The LLD figure is the percentage of initial rated lamp lumens which are still being emitted at the time of relamping. In order for this figure to be accurate, the expected time of relamping should be known, so that the suitable figure can be obtained from the data.

Luminaire dirt depreciation figures are suggested by manufacturers for various dirt conditions, but these figures are usually optimistic. The lighting designer can use either the mean LDD value, which means that the design level will be the average over the relamping period, or the end-of-relamping value, which means that the system will reach the design level only at the time the system is cleaned and relamped. Most indoor-design methods use the mean value.

Luminaire dirt depreciation varies with each situation. Even within one plant, there can be changes from one bay to the next, depending on the environment. The only way to ensure an accurate light-loss factor is to take in-plant measurements, use experience as a guide, or hire a lighting maintenance contractor to analyze a particular plant.

In most cases, however, the lighting designer must use the best possible estimate of light-loss factor in making calculations. This is usually LLD × LDD.

Area Area is given in square meters if the level of illumination is in lux, or square feet if the level of illumination is in footcandles.

POINT-BY-POINT CALCULATIONS

Point-by-point calculation procedures are used to determine the expected level of illumination at a particular horizontal or vertical point. This is often necessary when lighting is designed for areas which are long, narrow, and deep, such as warehouses and some hazardous locations. Chapter 7 covers design techniques for such areas, and also includes two charts which can be used to quickly determine the light level at a given point. One is for finding the light level at a point on a horizontal surface, and the other is for finding the light level on a vertical surface. The charts can be used for any lighting installation.

Bibliography

"A New Way to Calculate Plant Lighting Needs," *Plant Engineering,* October 1965, pp. 132–133.

Clark, Francis: "Accurate Maintenance Factors Part II," *Illuminating Engineering,* January 1966, pp. 37–46.

"The Determination of Illumination at a Point in Interior Spaces," LM-43, Committee on Lighting Design Practice of the Illuminating Engineering Society of North America: New York, 1974, ' reprinted from *Journal of the Illuminating Engineering Society,* January 1974.

Design Practice Committee of the Illuminating Engineering Society of North America: "Classification of Interior Luminaires by Distribution: Luminaire Spacing Criterion," *Lighting Design and Application,* August 1977, pp. 20–21.

Gaysunas, Ronald E.: "A Simplified Approach for Deriving Room Cavity Ratios," *Illuminating Engineering,* August 1964, pp. 553–554.

Jones, J. Roy, and Bill F. Jones: "Using the Zonal-Cavity System in Lighting Calculations," part I, *Illuminating Engineering,* May 1964, sec. II, pp. 413–415.

Jones, J. Roy, and Bill F. Jones: "Using the Zonal-Cavity System in Lighting Calculations," part II, *Illuminating Engineering,* June 1964, pp. 448–450.

Jones, Bill F., and J. Roy Jones: "Using the Zonal-Cavity System in Lighting Calculations," part III, *Illuminating Engineering,* July 1964, pp. 501–503.

Jones, Bill F., and J. Roy Jones: "Using the Zonal-Cavity System in Lighting Calculations," part IV, *Illuminating Engineering,* August 1964, pp. 556–561.

LeVere, Richard C., Robert E. Levin, and William C. Primrose: "Spacing Criteria for Interior Luminaires—The Practices and Pitfalls," *Journal of the Illuminating Engineering Society,* October 1973, pp. 41–49.

Lewin, Robert E.: "Revision of S/MH Concept," *Lighting Design and Application,* August 1977, pp. 22–25.

"Lighting Calculations," *IES Lighting Handbook,* 5th ed., sec. 9, Illuminating Engineering Society of North America, New York, 1972.

"Lighting Calculations for Interiors," *Fundamentals of Commercial and Industrial Lighting,* sec. 5, The Electrification Council, New York.

Lighting Design Practice Committee of the Illuminating Engineering Society of North America: "Calculations of Luminance Coefficients Based Upon the Zonal-Cavity Method," *Illuminating Engineering,* August 1968, pp. 423–432.

Lighting Design Practice Committee of the Illuminating Engineering Society of North America: "Zonal-Cavity Method of Calculating and Using Coefficients of Utilization," *Illuminating Engineering,* May 1964, sec. I, pp. 309–328.

O'Brien, Philip F.: "Lighting Calculations for Thirty-Five Thousand Rooms," *Illuminating Engineering,* April 1960, pp. 215–226.

5

Interior Lighting Design, Sample Problem and Analysis

There is no "typical" industrial space, but one sample area can be used to illustrate many of the factors which must be considered in the design of a satisfactory indoor industrial lighting system. This chapter covers the design procedures necessary to solve a lighting problem. The sample lighting system is first designed to average levels, and then a point-by-point analysis is made.

The background material used in the solution of the sample problem is given in Chap. 4, and only parts of it will be repeated in this chapter. The example is intended to show the design steps and decision points in the selection and evaluation of an interior lighting system. The designer should follow this process when planning a new installation or replacing an old one.

Analyzing the Space

Table 5-1 gives a suggested work sheet that can be used for an entire plant, if the conditions are similar, or a section of a plant. As the designer gathers data, information should be entered on such a work sheet; the

Table 5-1 Work sheet

Project Name ____________

Major Type of Work ____________

Maintained Illumination Level ________ Initial Level ________

ROOM DIMENSIONS:

Length ________ Width ________

Area ________

CAVITY DIMENSIONS: CAVITY RATIOS:

Ceiling-cavity height h_{cc} ________ CCR ________

Room-cavity height h_{rc} ________ RCR ________

Floor-cavity height h_{fc} ________ FCR ________

REFLECTANCES: EFFECTIVE CAVITY REFLECTANCES:

Ceiling ________ Ceiling ________

Walls ________

Floor ________ Floor ________

Lamp Type ________ LLD ________ Initial Lamp Lumens ________

Luminaire Type ________ SC ____ LDD ____ CU ____

LS-NC Rating ________ Ballast Factor ________

Ambient Temperature ____________

Ambient Dirt Condition ____________

Type of Hazardous Location ____________

sheet then becomes a handy reference when calculating luminaire quantities.

The first heading is the project identification, such as "XYZ Manufacturing Plant, Detroit, Michigan."

The second is the major type of work that is done or will be done in the area. This involves a detailed analysis of the space. Is the area to be used for more than one purpose, with perhaps more than one recommended light level? Is one section of the room to be used for a task which requires substantially more light than that provided by the general lighting? If so, the designer should follow the design procedures outlined in this chapter and then refer to the recommendations in Chap. 6, Supplementary Lighting for Specific Tasks, to ensure that each worker has adequate illumination.

It is also necessary to know the ambient conditions which will be present in the space, since this is a factor in luminaire choice. This information should be entered on the work sheet.

For the sample problem, assume that metalworking will be performed in the factory. It is a machine shop, and medium bench and machine work will be done. The interior environment will contain oil mist and residue from welding and grinding operations.

Recommended Level of Illumination

The recommended average maintained level of illumination can be found in Chap. 1. The light level recommended by the Illuminating Engineering Society of North America (IES) for medium bench and machine work is 110 dekalux (100 footcandles). For calculation purposes, the figure of 1076 lux will be used in place of 110 dekalux, since this is the exact equivalent of 100 footcandles.

Room Dimensions

The dimensions and area of the space to be lighted are noted in the next section of the work sheet. In industrial lighting, designing a lighting system as if the area were an empty room can produce poor results. If at all possible, the designer should obtain the plant layout for the area. This should include the placement of machinery, ducts, conveyors, etc., which could keep light from reaching the work plane. If the space is cluttered with tall machines or processing equipment, the area should be divided into smaller sections for design purposes.

For the sample problem, assume that the industrial facility is 61 by 36.6

meters (200 by 120 feet). The area of the space is 2232.6 square meters (24,000 square feet).

The room is divided into four bays but can be treated as a single room, since there are no tall pieces of machinery and the bays are not separated by partitions.

The height of the room cavity (h_{rc}) is 6.1 meters (20 feet). The height of the room cavity is used as the mounting height for industrial systems. The height of the ceiling cavity (h_{cc}) is 0.9 meters (3 feet), and the height of the floor cavity (h_{fc}) is also 0.9 meters.

Reflectances

Measured or estimated reflectances should be entered on the work sheet. For the example, assume that they are 50 percent for the ceiling, 30 percent for the walls, and 20 percent for the floor.

Generally, estimated ceiling and floor reflectances are used as the effective reflectances. Calculations for effective ceiling and floor cavity reflectances using the ceiling and floor cavity ratios are often ignored, since the actual surface reflectance values are seldom known with any degree of accuracy. There is also no way of determining the effects of such things as ducts, ceiling trusses, pipes, cables, etc., on the cavity reflectances. The nominal reflectance values used in the coefficient of utilization (CU) tables can be used with very little effect on the final results. Changes of 10 to 20 percent in room reflectance will change the average illumination level only slightly. Other factors such as the final luminaire quantity, spacing, and lamp lumens will have a much greater effect on the actual illumination level.

For the sample problem, then, assume an effective floor reflectance of 20 percent. Since there may be times when the designer wishes a more accurate ceiling reflectance figure, this will be derived for the sample problem.

For this derivation the cavity ratios of the area must be determined. To find the room cavity ratio (RCR), use the formula

$$\text{RCR} = \frac{5h(\text{room length} + \text{room width})}{\text{room length} \times \text{room width}}$$

In this case

$$\text{RCR} = \frac{5 \times 6.1 \times (61 + 36.6)}{61 \times 36.6} = \frac{2976.8}{2232.6} = 1.3$$

An alternate method for finding the RCR is to use the chart in Fig. 5-1. First find the room length of 61 meters on the lower ordinate of the left section of the graph. Find the spot between the 32-meter and 40-meter

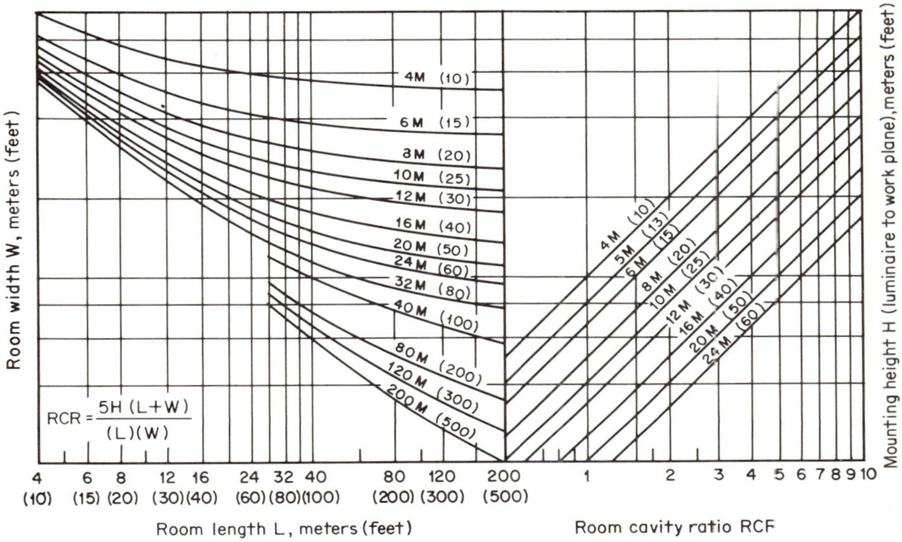

Fig. 5-1 Chart used to find RCRs.

lines which approximates the room width of 36.6 meters. Move directly right into the next section of the graph and meet the diagonal line representing the height of the room cavity, which in this case is 6.1 meters. Draw a line down to the bottom ordinate to find the RCR, which is 1.3.

The graph can be used with either meters or feet. Values on any one line are not equivalent.

In the sample problem, the ceiling cavity ratio (CCR) and the floor cavity ratio (FCR) are the same, since both cavities have the same dimensions. To find the ratios, use the same formula that was used for the RCR:

$$\frac{5h\,(\text{room length} + \text{room width})}{\text{room length} \times \text{room width}}$$

$$= \frac{5 \times 0.9 \times (61 + 36.6)}{61 \times 36.6} = \frac{439.2}{2232.6} = 0.2$$

The CCR and the FCR are both 0.2. Reading from Table 5-2, with an assumed ceiling reflectance of 50 percent, a wall reflectance of 30 percent, and a CCR of 0.2, the effective ceiling reflectance is 47 percent. A more complete table is included in the *IES Lighting Handbook.*

Selecting Lamp Type and Lumen Rating

At this point, the designer needs to make a tentative choice of lamp type and lumen rating. Assume the lamp type to be used is high-pressure sodium (HPS) because of its high efficacy.

Table 5-2 Percent effective ceiling and floor cavity reflectances and

Percent Ceiling or Floor Reflectance		90				80			
Percent Wall Reflectance		90	70	50	30	80	70	50	30
Ceiling or Floor Cavity Ratio	0	90	90	90	90	80	80	80	80
	0.1	90	89	88	87	79	79	78	78
	0.2	89	88	86	85	79	78	77	76
	0.3	89	87	85	83	78	77	75	74
	0.4	88	86	83	81	78	76	74	72
	0.5	88	85	81	78	77	75	73	70
	0.6	88	84	80	76	77	75	71	68
	0.7	88	83	78	74	76	74	70	66
	0.8	87	82	77	73	75	73	69	65
	0.9	87	81	76	71	75	72	68	63
	1.0	86	80	74	69	74	71	66	61

SOURCE: Reprinted with permission from *Illuminating Engineering,* April 1965 p. 281.

The wattage or lumen output of the lamp is going to determine the luminaire spacing, and consequently the ratio of spacing to mounting height (S/MH). For general industrial machine shops, luminaire spacing should approximate mounting height. As a very rough estimate, 50 percent of lamp lumens usually reach the work plane; initial lamp lumens can therefore be approximated with the formula

$$2 \times \text{maintained illumination level} \times MH^2$$

Since the mounting height is 6.1 meters, the formula for the example would read, $2 \times 1076 \times 6.1^2 = 80{,}076$ lumens. This is the initial rated lamp lumens that could be chosen for this application without exceeding the suggested S/MH ratio of 1:1. The same result is obtained if feet and footcandles are used in place of lux and meters: $2 \times 100 \times 20^2 = 80{,}000$ lumens.

The HPS lamp with an initial lamp lumen rating close to this number is the 400-watt lamp, at 50,000 lumens. This would be the logical choice. The 1000-watt HPS lamp is the next highest; it is rated at 140,000 lumens, which would probably prove too high for this application.

This is only a rough way of choosing a tentative lamp size. Once the design parameters are known, a lamp with the next higher or lower lumen rating may be called for. For the moment, however, assume that a 400-watt HPS source will be used.

Luminaire Selection

Next, a tentative luminaire choice should be made. Luminaire selection is based on a number of factors which must be considered almost simultaneously.

floor cavity correction factors for interior lighting calculations

70			50			30				10		
70	50	30	70	50	30	65	50	30	10	50	30	10
70	70	70	50	50	50	30	30	30	30	10	10	10
69	69	68	59	49	48	30	30	29	29	10	10	10
68	67	66	49	48	47	30	29	29	28	10	10	9
68	66	64	49	47	46	30	29	28	27	10	10	9
67	65	63	48	46	45	30	29	27	26	11	10	9
66	64	61	48	46	44	29	28	27	25	11	10	9
65	62	59	47	45	43	29	28	26	25	11	10	9
65	61	58	47	44	42	29	28	26	24	11	10	8
64	60	56	47	43	41	29	27	25	23	11	10	8
63	59	55	46	43	40	29	27	25	22	11	9	8
63	58	53	46	42	39	29	27	24	22	11	9	8

AMBIENT DIRT CONDITIONS

Since the work environment in the example has oil mist and residue in it, a sealed and filtered luminaire should be used for this application. The area contains no corrosive fumes or moisture that would require specially adapted luminaires. The area is not classified as hazardous.

AMBIENT TEMPERATURE

The temperature surrounding the luminaires will not exceed 38°C (100°F); a ballast temperature rating of 40°C or higher can therefore be used.

NOISE CRITERION

The expected noise criterion level for a machine shop is 45 to 70 dB, as noted in Table 3-4, Chap. 3. Therefore, the lighting system noise criterion (LS-NC) rating of the luminaire should be 45 or less in order for the luminaire to be inaudible. The LS-NC rating for the proposed luminaire choice is 42.

MOUNTING HEIGHT

The mounting height in this case is 6.1 meters. Either a luminaire designed for low mounting heights or a standard luminaire can be used. For this example, a standard industrial luminaire will be chosen, since this type tends to be more efficient in intermediate or high bay applications.

BEAM SPREAD

The particular 400-watt HPS luminaire that will be used for the example is available in a variety of beam spreads having spacing criterion (SC) values of 0.7, 1.0, 1.3, and 1.5. Distribution data for each of these luminaires is given in Table 5-3. The figures are expressed per 1000 lamp lumens.

As described in the previous chapter, for best results each luminaire should provide no more than 50 percent of the illumination directly beneath it. This means that the nadir candlepower, or candelas at 0 degrees, should be close to the value required to produce this level.

The maximum nadir candelas for this application can be determined with the following formula:

$$\text{max cd} = E \times \text{contribution of each luminaire} \times \text{MH}^2$$

where max cd = maximum nadir candelas
E = maintained or initial illumination level of 1076 lux
contribution each luminaire = 50 percent, or 0.5
MH = mounting height of 6.1 meters

Substituting in the formula, $1076 \times 0.5 \times 6.1^2 = 20{,}019$ candelas. If footcandles and feet are used, the same formula applies: $100 \times 0.5 \times 20^2 = 20{,}000$ candelas.

Nadir candlepower of the luminaires being considered can be read from their distribution data. The figures in the data are expressed per 1000 lumens; they must be multiplied by actual lamp lumens. The 400-watt HPS lamp is rated at 50,000 lumens, and so the nadir candelas in each case should be multiplied by 50.

The candelas at 0 degrees are as follows:

0.7-SC luminaire: 39,505 candelas
1.0-SC luminaire: 27,980 candelas
1.3-SC luminaire: 22,500 candelas
1.5-SC luminaire: 17,120 candelas

For this application the luminaire with the 1.3 SC was chosen, since it emits 22,500 candelas at 0 degrees. The luminaire with the 1.5 SC would also be acceptable.

SHIELDING ANGLE AND REFLECTOR SHAPE

Photometric data for the luminaire with the 1.3 SC indicates that direct light from the luminaire is shielded at 45 degrees below the horizontal. Direct high-intensity discharge (HID) luminaires should have a shielding angle greater than 35 degrees; this reflector will thus provide a low-brightness system.

The luminaire comes in two reflector shapes when used with a 400-watt source. The beam-spread choice determines the reflector shape, since the range of beam spreads is created by changes in socket position, reflector contour, and reflector finish.

Table 5-3 Distribution data (per 1000 lumens)

Vertical Angle	0.7 SC	1.0 SC	1.3 SC	1.5 SC
Degrees	Average Candelas			
180	0	3.1	3.2	3.2
175	0.11	3.8	4	3.3
165	0.15	3.2	2.8	1.7
155	13	4.1	8.1	11.2
145	18.3	24.2	26.7	24.4
135	8.9	32.8	26.3	17.5
125	1.4	27	17.8	7
115	0.9	18.2	6.9	1.4
105	0.9	8.3	1.5	0.9
95	0.7	1.7	1.3	0.8
85	0.6	1.1	1.2	0.8
75	0.9	1.3	1.3	1.1
65	0.65	1.5	1.4	1.5
55	30.3	1.9	3	19.5
45	94	65.1	134	212.3
35	208.8	296.9	346.2	365.5
25	454.3	412.5	416.1	382.6
15	820.5	577.2	460.1	325.7
5	773.4	562.8	443.8	325
0	790.1	559.6	450	342.4
	Zonal Lumens			
170–180	0.02	0.37	0.37	0.3
160–170	0.04	0.9	0.8	0.5
150–160	6	1.9	3.8	5.2
140–150	11.5	15.2	16.8	15.3
130–140	6.9	25.4	20.4	13.6
120–130	1.2	24.2	16	6.3
110–120	0.9	18	6.9	1.4
100–110	0.9	8.8	1.6	1
90–100	0.7	1.8	1.4	0.9
80–90	0.7	1.2	1.3	0.9
70–80	0.9	1.3	1.3	1.1
60–70	0.65	1.5	1.4	1.5
50–60	27.2	1.7	2.7	17.5
40–50	72.8	50.4	103.7	164.3
30–40	131.1	186.4	217.4	229.6
20–30	210.3	191	192.7	177.2
10–20	232.2	163.4	130.2	92.2
0–10	73.5	53.5	42.2	30.9

STANDBY LIGHTING

Because of the moving machinery in this plant, it needs a standby system of lighting that will come on instantly when power is restored after a momentary power interruption. Any HID luminaire can be equipped with a quartz lamp inside the reflector. This standby lighting should be specified for about 10 percent of the luminaires.

Coefficient of Utilization

Next, the CU should be determined for this application. Table 5-4 gives the CU table for the luminaire with the 1.3 SC. The effective reflectances and the RCR for the sample problem can be used to check the CU. These figures can be found on the work sheet.

An effective ceiling cavity reflectance of 47 percent (rounded to 50 percent), a 30 percent wall reflectance, and a 20 percent floor cavity reflectance should be used. The RCR is 1.3. Table 5-4 shows that with these reflectances and an RCR of 1, the CU is 0.742. For an RCR of 2, the CU is 0.690. The CU for 1.3 falls between these two values, and must be interpolated by subtracting 0.3 times the difference between the two CUs from the CU for the RCR of 1. Thus:

$$0.742 - 0.690 = 0.052$$
$$0.052 \times 0.3 = 0.016$$
$$0.742 - 0.016 = 0.726$$

The CU for an RCR of 1.3, and for the given floor, wall, and ceiling reflectances, is therefore 0.726.

Light-Loss Factors

Once the lamp and luminaire choices are made, the value for the lamp lumen depreciation (LLD) can be read from the manufacturer's data. The published mean LLD for the 400-watt HPS lamp is 0.9.

Luminaire dirt depreciation (LDD) can be read from charts published by the luminaire manufacturer, such as the one in Table 5-5. The LDD figures can also be obtained from previous experience with similar luminaires or from a lighting-maintenance company. For a sealed and filtered luminaire in a medium dirt condition, the LDD would be approximately 0.93.

Ballast factor (BF) is not normally used with HID lighting, but variations in lamp voltage and ballast characteristics can sometimes result in a lamp producing other than its rated lumens. The ballast factor is multi-

Table 5-4 Coefficients of utilization—zonal-cavity method

	Effective Ceiling Cavity Reflectance–Percent															
	80			70			50			30			10			0
Wall Reflectance	50	30	10	50	30	10	50	30	10	50	30	10	50	30	10	0
Room Cavity Ratio (RCR) 0																
1	.825	.802	.788	.802	.783	.769	.758	.742	.734	.719	.707	.700	.681	.671	.666	.648
2	.763	.734	.708	.748	.719	.698	.710	.690	.669	.678	.662	.646	.647	.636	.623	.607
3	.709	.674	.644	.695	.661	.636	.666	.638	.615	.641	.618	.599	.614	.595	.578	.564
4	.663	.619	.589	.642	.609	.578	.625	.592	.563	.603	.573	.554	.578	.556	.539	.526
5	.612	.570	.539	.601	.561	.532	.582	.549	.522	.561	.533	.511	.544	.519	.500	.486
6	.571	.527	.496	.564	.522	.487	.542	.509	.482	.528	.497	.470	.512	.485	.462	.450
7	.532	.485	.450	.513	.476	.446	.506	.466	.439	.490	.455	.431	.477	.446	.425	.414
8	.491	.443	.413	.473	.437	.408	.469	.434	.405	.455	.421	.396	.442	.410	.391	.379
9	.453	.408	.374	.437	.401	.372	.432	.391	.364	.420	.387	.359	.412	.381	.354	.344
10	.420	.370	.340	.389	.368	.339	.400	.359	.333	.391	.354	.328	.381	.350	.323	.313

Basis of calculations:
Ratio of spacing to mounting height .7
Effective floor cavity reflectance 20% ($P_{FC} = 0.20$)

Table 5-5 Luminaire dirt depreciation factors

	Luminaire Dirt Depreciation		
Luminaire Type	Light	Medium	Heavy
Sealed and filtered	0.97	0.93	0.88
Enclosed	0.94	0.86	0.77
Open and ventilated	0.94	0.84	0.74

plied by the published initial rated lumens of the lamp to obtain the lumens actually generated in service.

For the example, assume a BF of 100 percent, or 1. The light-loss factor in this case is thus

$$\text{LLD} \times \text{LDD} \times \text{BF} = 0.9 \times 0.93 \times 1 = 0.837$$

In order for 1076 lux (100 footcandles) to be the maintained light level, the initial level must be 1076/0.837 = 1285 lux (100/0.837 = 119 footcandles).

Number of Luminaires

By now, the designer should have a completed work sheet and should refer to it for the figures necessary to compute luminaire quantities. Table 5-6 gives the completed work sheet for the sample problem.

To find the number of luminaires for this application, use the following formula:

$$N = \frac{E \times \text{area}}{\text{LL} \times \text{CU} \times \text{LLD} \times \text{LDD} \times \text{BF}}$$

where N = number of luminaires
E = maintained level of illumination
area = area of the space to be lighted
LL = initial rated lamp lumens
CU = coefficient of utilization
LLD = lamp lumen depreciation
LDD = luminaire dirt depreciation
BF = ballast factor

(The last three quantities could be combined into one light-loss factor.)

Substituting in the formula,

$$N = \frac{1076 \times 2232.6}{50{,}000 \times 0.726 \times 0.9 \times 0.93 \times 1} = \frac{2{,}402{,}277.6}{30{,}383.1} = 79$$

Table 5-6 Work sheet

Project Name sample problem

Major Type of Work metalworking

Maintained Illumination Level 1076 lux (100 footcandles) Initial Level 1285 lux (119 footcandles)

ROOM DIMENSIONS:

Length 61 meters (200 feet) Width 36.6 meters (120 feet)

Area 2,232.6 square meters (24,000 square feet)

CAVITY DIMENSIONS:		CAVITY RATIOS:
Ceiling cavity-height	h_{cc} 0.9 meters (3 feet)	CCR 0.2
Room cavity-height	h_{rc} 6.1 meters (20 feet)	RCR 1.3
Floor cavity-height	h_{fc} 0.9 meters (3 feet)	FCR 0.2

REFLECTANCES:	EFFECTIVE CAVITY REFLECTANCES:
Ceiling 50%	Ceiling 47%
Walls 30%	
Floor 20%	Floor 20%

Lamp Type HPS 400 watt LLD 0.9 Initial Lamp Lumens 50,000

Luminaire Type filtered SC 1.3 LDD 0.93 CU 0.726

LS-NC Rating 42 Ballast Factor 1.0

Ambient Temperature below 38°C

Ambient Dirt Condition medium

Type of Hazardous Location none

The resulting luminaire quantity is a prime number. In order to divide the luminaire quantity into a number of rows containing an equal number of luminaires, the quantity will have to be increased to 80.

The same results would be obtained if feet and footcandles were used in the formula.

The area covered by each of 80 luminaires would be 2232.6/80 = 28 square meters, or 24,000/80 = 300 square feet.

Spacing

Square spacing is desirable, but it is not required in a rectangular space. To determine the square spacing for the example, use the formula

$$\sqrt{\frac{\text{area}}{N}} = \sqrt{\frac{2232.6}{80}} = 5.3 \text{ meters} \quad \text{or} \quad \sqrt{\frac{24{,}000}{80}} = 17.3 \text{ feet}$$

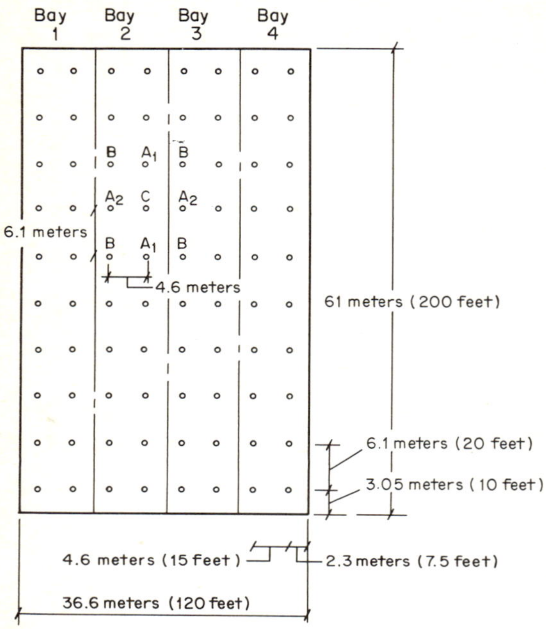

Fig. 5-2 Final layout for the sample problem. There are eight rows of 10 luminaires each, spaced at 4.6 meters in one direction and 6.1 meters in the other. Nine luminaires have been picked at random and labeled. Illumination under each luminaire should come partially from the luminaire overhead and partially from adjacent luminaires.

This spacing does not fit either room dimension evenly. In order to keep a minimum of two rows in each bay, the spacing across the bay should be 4.6 meters (15 feet) for 80 luminaires. The spacing in the other direction is 6.1 meters (20 feet).

Figure 5-2 is a diagram of the final spacing for the sample problem. The luminaires near the walls have been placed at a distance from the wall equal to half the spacing in that direction. Thus, a luminaire in a corner of the room is 2.3 meters (7.5 feet) from one corner wall and 3.05 meters (10 feet) from the other.

Finding Levels of Illumination

The graph in Fig. 5-3 can be used to plot the actual levels of illumination throughout the space. From this data, the relative contribution of each luminaire to a particular point can be determined. Relative contribution can also be determined by using the expanded SC graph in Chap. 4, but the SC graph does not give actual levels of illumination.

Figure 5-3 is based on the equation

$$E = \frac{\text{cd} \cos^3 \theta}{\text{MH}^2}$$

where E = actual level of illumination
cd = candela value given in photometric curve
$\cos^3 \theta$ = cosine cubed of incident angle
MH = mounting height of the luminaire

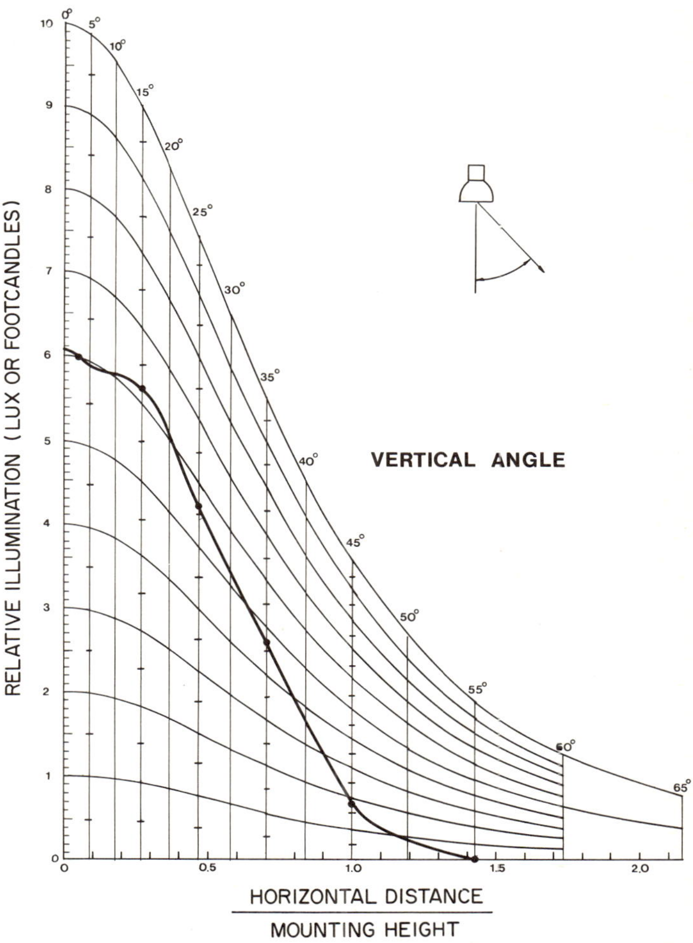

Fig. 5-3 Chart used to calculate actual illumination levels and relative contributions of luminaires to the illumination at a point. Plots of distribution data can be redrawn to the scale of working drawings and used as templates.

Candela values can be expressed in photometric data on the basis of actual lamp lumens or per 1000 lamp lumens. The illumination level (E) can be either an initial or a maintained level. If the mounting height is expressed in meters, the illumination level will be in lux; if it is expressed in feet, the level of illumination will be in footcandles.

When the formula is used, a value for actual illumination levels is obtained. When using the graph, the left ordinate can be any number of lux or footcandles. To convert the relative values read from the graph to actual levels of illumination, a constant (K) is added to the original formula, and it is rewritten

$$E = \text{cd} \times K \times (\cos^3 \theta)$$

The value of K is determined by the formula

$$K = \frac{\text{LF} \times \text{LLF}}{\text{MH}^2}$$

where $\text{LF} = \dfrac{\text{lamp lumens used in system}}{\text{lamp lumens used in photometric data}}$

LLF = light-loss factor, used only if the plot is based on the maintained level of illumination

MH = mounting height of the luminaire above the work plane

LEVELS OF ILLUMINATION FOR THE SAMPLE PROBLEM

The HPS luminaire with an SC of 1.3 was used to light the sample industrial area. The distribution data for this luminaire is given in Table 5-3. In order to plot the distribution data on Figure 5-3, follow this procedure:

Step 1. Determine the constant K. In this case, distribution data is expressed per 1000 lamp lumens. Since the lamp is rated for 50,000 initial lamp lumens, the lamp factor (LF) is 50,000/1000 = 50.

Initial rather than maintained illumination levels will be considered, so the light-loss factor (LLF) is 1.

Mounting height is 6.1 meters (20 feet).

The constant is thus $50 \times 1/6.1^2 = 50/37.21 = 1.3437$.

Step 2. Find the illumination level at each vertical angle by multiplying the candela value from Table 5-3 by the constant 1.3437, since the formula is $E = \text{cd}\ K \cos^3 \theta$ and the cosine is taken account of in the graph. Plot these values on the graph by locating each value on the left ordinate and following the curved line until it reaches the vertical line indicating the appropriate angle. The angles are noted at the top of the graph. For example, at 0 degrees this luminaire emits 450 candelas. Since $450 \times 1.3437 = 604.7$, the first point is on the 0-degree line at 605 lux.

Connect the points obtained for each vertical angle with a smooth curve.

Step 3. The final curve represents the level of illumination at any point away from the point on the work plane directly below the luminaire. The illumination values can be read directly from the left side of the graph. If desired, photometric data for more than one luminaire can be plotted on the graph so that the level of illumination provided by one luminaire can be compared directly to the level provided by another.

Step 4. If desired, replot the data to the scale of the working drawings. The bottom line of the graph is expressed in ratios of the horizontal distance to the mounting height. The actual distances are determined by multiplying each value by the mounting height. Replotted curves can be used as templates to determine the level of illumination at any point in the lighted area.

Step 5. Determine vertical illumination values on a plane perpendicular to a line drawn from the luminaire to the point being measured. This can be done by multiplying the horizontal illumination values by the ratio of horizontal distance to mounting height. For instance, the horizontal illumination at a distance from the luminaire of 0.6 times the mounting height is approximately 340 lux. The vertical illumination in line with the luminaire is thus $0.6 \times 340 = 204$ lux.

RELATIVE CONTRIBUTION OF EACH LUMINAIRE

Nine luminaires from the area used as an example have been noted on Fig. 5-2. In an industrial area, the luminaire C should contribute about 50 percent of the illumination directly beneath it. The graph in Fig. 5-3 can be used to check relative contributions.

The C luminaire contributes 605 lux at the nadir or directly beneath it. The A_1 luminaires are 6.1 meters, or 1 mounting height, from the C luminaires. The ratio of horizontal distance to mounting height is thus 1:1. Find the 1.0 point on the lower ordinate and go up to the plotted line. Crossing over to the left ordinate, note that this luminaire contributes 60 lux to the area under C. There are two A_1 luminaires; the total contribution is therefore $2 \times 60 = 120$ lux.

The A_2 luminaires are 4.6 meters from luminaire C, so the horizontal distance divided by the mounting height is $4.6/6.1 = 0.75$. Each of the A_2 luminaires contributes 200 lux, for a total of $2 \times 200 = 400$ lux.

The B luminaires will make a negligible contribution.

The total contribution to the area under C is 605 + 120 + 400 = 1125 lux.

The C luminaire contributes 605/1125 = 0.54, or 54 percent of the illumination to the area directly beneath it. The four A luminaires contribute the other 46 percent. This is a good relationship for an industrial facility, where light may be blocked by machinery, conveyors, workers' shadows, etc.

ILLUMINATION AT A POINT

The level of illumination at the point directly under luminaire C will be slightly higher than the calculated value of 1125 lux because of some contribution from other luminaires in the factory and the illumination produced by light reflected off the room surfaces. Light levels will probably come close to the recommended initial value of 1285 lux.

Other points in the area can be checked in the same manner, by using Fig. 5-3. This is particularly important if a minimum level of illumination has been specified.

Point-by-Point Analysis

Figure 5-2, the layout of the sample area, is divided into four bays, each 9.1 by 61 meters (30 by 200 feet). There are two rows of luminaires in each bay. A point-by-point analysis was made of the direct illumination levels provided throughout the space.

The two rows of luminaires in a bay next to the wall, bay number 1, were then presumably turned off, and an analysis was made of the lighting in the adjoining bay, number 2.

These evaluations were done by computer. They can also be done with the data obtained from Fig. 5-3. Computerized design methods are used by many manufacturing and engineering firms.

CROWNING EFFECT

Figure 5-4 shows the results of this analysis. Luminaires whose light patterns overlap will cause a crowning of the illumination level in the center of the area being lighted. This is more pronounced in smaller rooms with higher RCRs. The RCR for a 9.1- by 61-meter bay is 3.8.

Generally, the first row of luminaires in an area is placed at a distance from the wall equal to one-half the luminaire spacing. When this is done, the light level does not reach a peak until the second row of luminaires, as shown in Fig. 5-4. The only way to avoid this is to use luminaires that have a more concentrating, narrower beam spread and to start them closer to the walls. This is inefficient, and probably unnecessary, if most activity is located away from outside walls.

Turning Off One Bay of Lights The crowning effect becomes more pronounced when lights in adjacent bays are turned off. If there is no light in the first bay, the illumination at the edge of the second bay can drop 30 to 50 percent from the level in the center of the room. When the lighting in the third bay is also turned off, the crowning effect within the second bay becomes obvious (Fig. 5-4).

Task-Oriented Lighting It is important to understand the crowning effect when designing task-oriented lighting. If one area is to receive a higher level of illumination than another, the first luminaire should be located at or beyond the perimeter of the lighted area, in order to compensate for the crowning effect. If this is not done there will be a significant change in illumination between the perimeter and the center of the area. Illumination at each point should be provided by at least four luminaires, particularly at low mounting heights.

Variations in Mounting Height Variations in mounting height within a plant do not change the utilization of the system except at the perimeter. When the ceiling is peaked, the spacing can be kept the same and the luminaires can follow the ceiling. The average ceiling height can be used as the mounting height for calculation purposes.

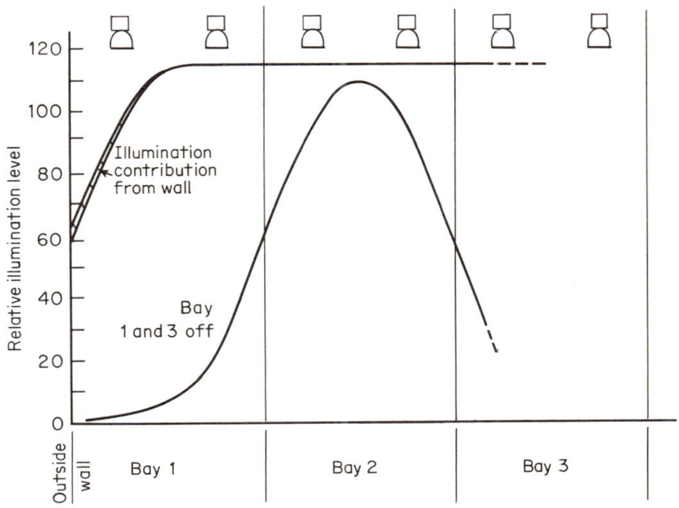

Fig. 5-4 Illustration of crowning effect. Illumination tends to drop off near the perimeter of an area, particularly if the RCR is high, as in a small room. Note how little contribution is made by light reflected off the wall.

Fig. 5-5 Walls do not reflect light uniformly, as illustrated here.

There will be more overlap of beams if the ceiling is peaked, but the only loss in light level will be at the perimeter of the area.

Clerestory or Monitor Construction In clerestory and monitor construction, losses on the upper side walls can be reduced by using luminaires with narrower beams and by spacing luminaires away from the walls. The added mounting height can be used to increase the luminaire spacing without changing the S/MH ratio. The only area that will lose light will be the perimeter of the plant, where the added mounting height will increase the losses due to light striking the end walls.

WALL CONTRIBUTION

The shaded area in Fig. 5-4 represents the contribution of light reflected off the wall. Reflected light makes only a minor contribution toward improving the illumination along the perimeter of an industrial plant. In the example, with a 6.1-meter (20-foot) room cavity, the light reflected off the walls increases the illumination by less than 10 percent.

Light from a luminaire never lights the perimeter walls uniformly. The photograph in Fig. 5-5 shows the inside of an industrial facility. Note how the brightness of the walls varies.

A computer was used to calculate the vertical illumination on various sections of the wall along the first bay in the sample problem. The approach used assumes that the wall is a diffuse reflector, that is, that wall

Table 5-7 Configuration factors

Vertical Section of Wall	Distance from Wall: 0–1.5 meters (0–5 feet)	1.5–3 meters (5–10 feet)	3–4.6 meters (10–15 feet)	4.6–6.1 meters (15–20 feet)	6.1–7.6 meters (20–25 feet)
0–1.5 meters (0–5 feet)	0.148	0.028	0.012	0.008	0.004
1.5–3 meters (5–10 feet)	0.192	0.120	0.06	0.036	0.024
3–4.6 meters (10–15 feet)	0.06	0.092	0.076	0.052	0.036
4.6–6.1 meters (15–20 feet)	0.026	0.06	0.062	0.052	0.046
6.1–7.6 meters (20–25 feet)	0.018	0.04	0.05	0.042	0.038

brightness does not change with the direction of the incident illumination. The approach also assumes that the wall's length is very long in comparison to its height (at least 10 times as long), as is the case in the sample problem.

The wall and work plane were divided into 1.5-meter (5-foot) sections to improve the accuracy of the results, since the wall is not uniformly lighted. This is shown in Fig. 5-6. If the different sections have different reflectivities, these can also be considered.

Each zone is assumed to be uniformly lighted. Average illumination values are calculated for the midpoint of each zone.

Table 5-7 gives configuration factors which are used to calculate the

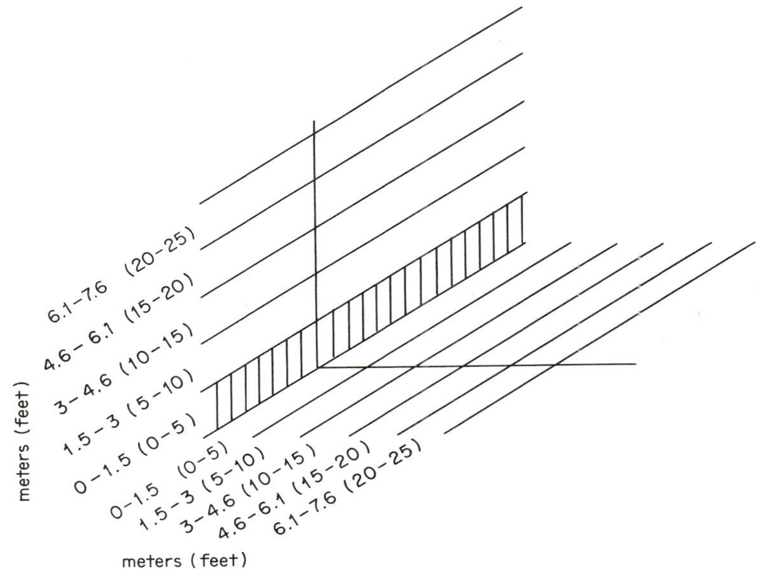

Fig. 5-6 The work plane and the area of the wall above the work plane are here divided into 1.5-meter zones.

Table 5-8 Illumination contribution from wall to work plane (from wall with 30% reflectance)

Vertical Section of Wall, Meters*	Vertical Illumination On Wall, Lux†	Work Plane: Distance from Wall in Meters				
		0–1.5	1.5–3	3–4.6	4.6–6.1	6.1–7.6
0–1.5	282	12.5	2.4	1.0	0.7	0.3
1.5–3	363	20.9	13.1	6.45	3.9	2.6
3–4.6	199	3.6	5.5	4.5	3.1	2.2
4.6–6.1	13	0.02	0.15	0.18	0.15	0.15
Totals, Lux		37.02	21.15	12.13	7.85	5.25

*1 meter = 3.3 feet.
†1 lux = 0.09 footcandles.

illumination on the work plane generated by light reflected off each section of the vertical wall. To calculate levels of illumination use the following formula:

Illumination on work plane = incident illumination on wall × wall reflectance × configuration factor

The vertical incident illumination on the wall is included in Table 5-8.

The average illumination of the first zone of the wall, which extends from 0 to 1.5 meters (0 to 5 feet) is 282 lux. The wall reflectance for the sample problem is 30 percent. The configuration factor for the contribution of this section to the first 1.5 meters of work plane adjacent to the wall is 0.148, as read from Table 5-7. Substituting in the formula,

$$E = 282 \times 0.3 \times 0.148 = 12.5 \text{ lux.}$$

All of the figures in Table 5-8 were derived in this manner.

Light reflected from the wall does not make any significant contribution to the level of illumination on the work plane. The section next to the wall receives a total of 37 lux of reflected light, but since the average illumination level is 1285 lux, this is only 37/1285, or 2.9 percent of the total.

In industrial facilities, although it is desirable to paint walls with light colors, their contribution is minimal. The same is true of the ceiling. High-intensity discharge luminaires characteristically radiate less than 10 percent of their light above 90 degrees. A lighter ceiling reduces the contrast between luminaires and the ceiling and also makes the ceiling brightness as high as the floor brightness. From the standpoint of appearance this is desirable. But unless the plant is very clean, high ceiling reflectivity is difficult to maintain and cannot be relied on to play a major role in providing illumination.

As industrial areas get smaller, however, and mounting heights lower, the reflectance of room surfaces plays a greater role. In small offices or laboratories, reflectivity values and fresh paint are much more significant, and light levels can be influenced to a much greater extent. This is also true in industrial areas where visibility of metal surfaces is a factor.

6

Supplementary Lighting for Specific Tasks

General lighting may not be sufficient for very difficult visual tasks, such as fine mill work or inspection. In such cases, some sort of task-oriented supplementary lighting can be installed as well.

Supplementary lighting is used to improve task brightness and contrast. This can sometimes be done by mounting a low-wattage incandescent or high-intensity discharge (HID) luminaire directly on a machine and adjusting it empirically. Supplementary lighting can also involve a lighting system specifically designed beforehand to make the detail on a particular task more visible.

In either case, the designer must be familiar with the factors involved in vision, as well as the effects of the placement of a light source. This chapter includes background information needed before designing any type of supplementary lighting system. An analysis of different types of inspection systems is also included.

Types of Light Sources

Historically, lighting systems have been identified with the light source they use, for example, incandescent lighting, fluorescent lighting, or HID lighting. This tends to emphasize the way the light is generated rather than the characteristics of the lighting system itself.

Of greater importance in supplementary lighting is the effect which the light produces on the work surface. In this respect, lighting systems can be defined as point, linear, or area sources.

A *point source* produces a circular area of high luminance on a specular surface. Any bare incandescent lamp, reflector spotlamp, or HID source can serve as a point source.

A *linear source* produces a line of high luminance across a specular surface. The effect of a fluorescent lamp on a surface is that of a linear source.

A large area diffuse light source produces uniform luminance over a specular surface. This type of light source is generally produced by backlighting a plastic or glass diffuser. If the lamp is mounted behind a diffusing panel, it makes little difference what type of lamp is used. It is only when the lamp is exposed directly to the work area that the characteristics of the lamp itself become important.

Most lighting systems can be categorized somewhere between point sources and area or diffuse sources. Comparison can be made with outdoor lighting. On a clear, sunny day approximately 80 percent of the light comes directly from the sun, which can be considered a point source. The remaining 20 percent comes from the illumination of the sky, which forms a hemisphere of approximately uniform brightness and would be classified as an almost ideal diffuse light source. A point source is capable of producing distinct shadows, whereas a completely diffuse, uniformly lighted hemisphere is incapable of producing any distinct shadows.

A further consideration is the direction from which the light is coming. This is important with a point source but becomes less so as the lighting system becomes more and more diffusing. An extreme example of a directional point source is the lighting which is produced by the sun on the moon. The absence of any atmosphere creates extremely deep shadows. The only thing approaching this is a floodlight system using a single luminaire. Indoors, reflections off the ceiling, walls, and equipment within the space always add some diffuse component to the lighting.

EFFECT OF BACKGROUND

A point source or linear source can be backed by a light-reflecting surface to produce an effect midway between a point source and a diffuse source. This will tend to reduce the contrast produced by the reflected image of the light source but will also ensure that some reflected brightness will occur on the surface of the task between the images of the bare lamps.

Painting the ceiling of an industrial plant white gives a white background to the lighting and produces this effect. On a smaller scale, the lamps can be backed by a reflector or flat sheet of light-colored material. The net result is similar to what can be observed outdoors with the combination of sun and sky brightness, as opposed to the high contrast and deep shadows that are produced on the moon, where there is no diffuse component to the lighting.

Figure 6-1 shows the difference that a change in background can make in inspection lighting. In this example a fluorescent lamp was used. Gray steel is semispecular and fairly diffuse. The irregular surface defects are emphasized by the use of bare fluorescent lamps in contrast with a dark background. The high contrast and maximum luminance gradient between lamps makes it easier to see the defect in the gray steel.

However, the defects in both the specular aluminum and the black paint are easier to see under bare fluorescent lamps with a white background. The luminance gradient between lamps is less than with the black background but enough so that the defects in the material can be seen. Too large a black area between the lamp images makes it possible to lose some of the defects in the black paint and partially obscures defects in the specular aluminum. As a rule, the darker the surface and the more specular the finish, the greater the brightness that is required between the light sources to create some surface brightness between lamp images.

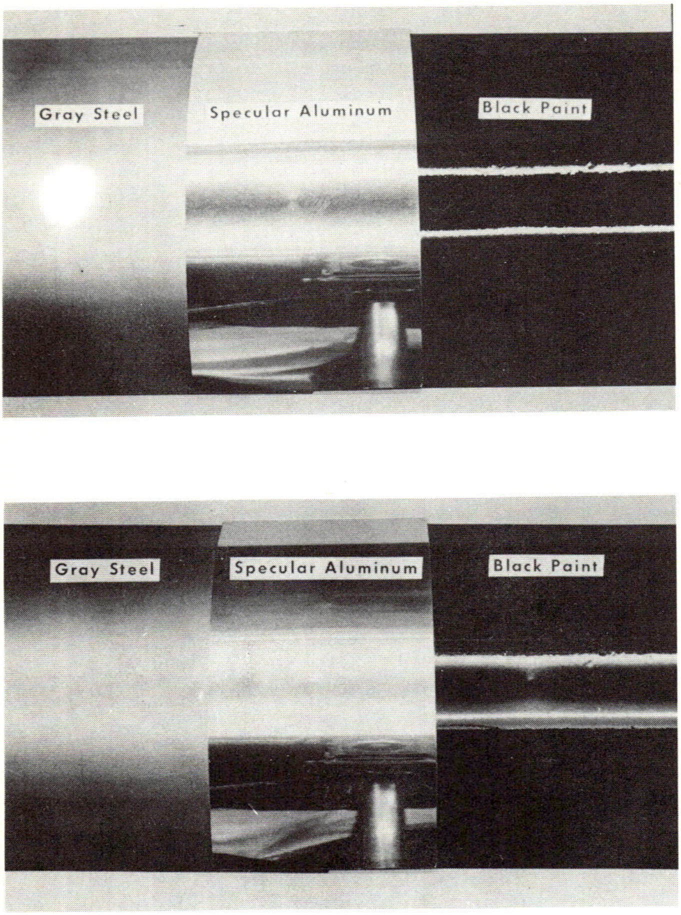

Fig. 6-1 The defect in the gray steel is more readily visible with bare lamps and a black background. However, adding a white background to the lamps makes the defects in specular aluminum and black paint easier to see.

AMOUNT OF LIGHT PRODUCED

The amount of light which a lamp or luminaire produces is proportional to its brightness and its area. There are wide differences in brightness between different light sources. A 400-watt high-pressure sodium (HPS) lamp, for instance, has an arc tube which is approximately 12.7 centimeters (5 inches) long and 0.95 centimeter (⅜ inch) in diameter. It produces almost 3 times as much light as a 1500-milliamp fluorescent lamp which is 2.4 meters (8 feet) long and 3.8 centimeters (1½ inches) in diameter. To reduce the brightness of the lighting system, therefore, it is necessary to increase the size of the luminaires so that the average brightness is reduced.

The type of light source which is used will affect the appearance of everything in the room. Replacing a fluorescent system with an HID system will change the visibility and the appearance of the task, even though the illumination level is the same. It is beneficial, therefore, to know something about the effect of different types of lighting systems, to ensure that the type of work and its detail can be seen.

So little energy is involved in supplementary lighting that it is not necessary to choose the most efficient source—just the one that is best suited to solving the specific problem involved.

Size of the Light Source

The size of a diffuse large-area light source is important in inspection lighting, lighting for machining of small metal parts, and other critical tasks. Figure 6-2 shows the procedure to be used in establishing the size of the source necessary to obtain the desired area of reflected brightness in a plain specular surface. It is based on the premise that the angle subtended by the source on the work is equal to the angle subtended by the luminous area at the eye; this principle is expressed by the equation $S = E$.

Figure 6-2 also shows the procedure to be used in establishing the size of the source necessary to obtain the desired area of reflected brightness on a convex specular surface. In this case

$$S = 2C + E,$$

where S = angle subtended by source on work
C = angle subtended by luminous area on work
E = angle subtended by luminous area at eye

Reflecting Characteristics of Surfaces

Another thing that the designer must analyze when designing a supplementary system is the type of surface and how it reflects light. For instance, metallic and nonmetallic surfaces have different characteristics. Light is reflected directly off a metal surface, whereas it penetrates a nonmetallic or dielectric surface, where it is diffused. Diffusion which occurs on a metal surface is due largely to the roughness of the surface rather than light being diffused within the surface itself. The diffusion which

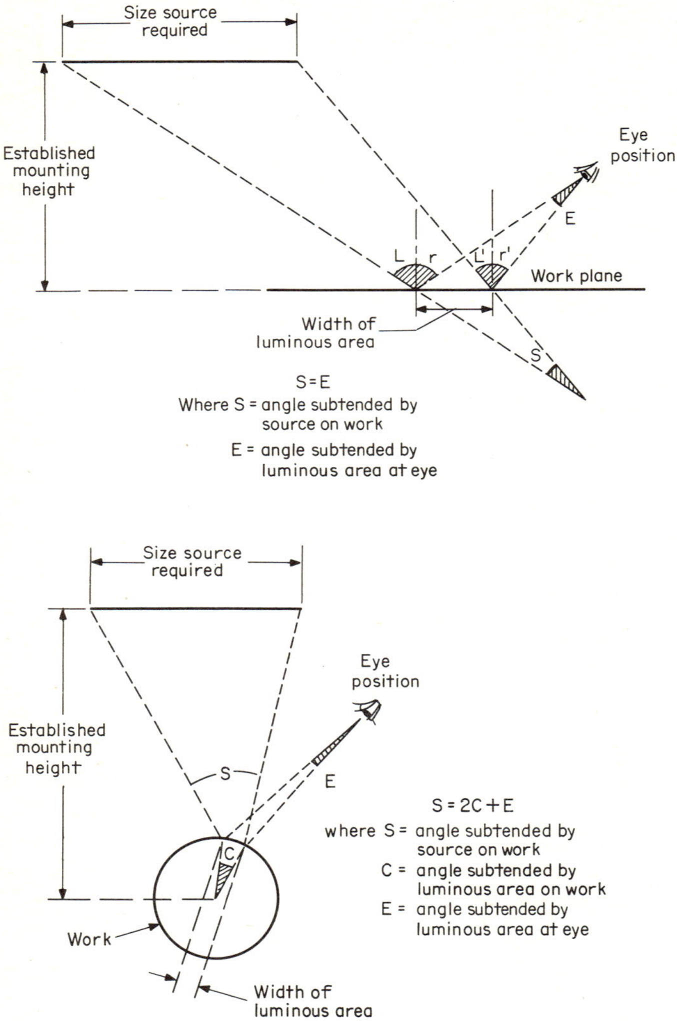

Fig. 6-2 Procedure to be used to establish the size of the source necessary to obtain the desired area of reflected brightness on a flat specular surface or on a convex specular surface. (These drawings are reproduced with permission from *Lighting for Machining of Small Metal Parts,* IES File No. 31/13, 1949.)

occurs within nonmetallic surfaces is the main factor which controls their appearance.

There are hundreds of different types of diffusing nonmetallic surfaces, whereas all metal surfaces tend to look somewhat the same. The main differences in appearance between different metal surfaces are caused by differences in roughness and, to a limited extent, in color. The higher the polish on a metal surface, the more it takes on the characteristics of a mirror. This can create problems when the brightness of the polished surface approaches the brightness of the luminaire. In this case the brightness of a luminaire will be a greater factor in visibility than the illumination level it produces. The situation will probably be glaring.

Levels of Illumination

The illumination level for machining of small metal parts, or other difficult industrial tasks, can be chosen by referring to appropriate charts of recommended levels. The Illuminating Engineering Society of North America (IES) publishes such charts, as shown in Chap. 1.

Some industrial tasks require as much as 10,760 lux (1000 footcandles). This would be for extrafine bench or machine work, grinding, or some other type of fine work that involves low contrast on a gray or dark background. It would be expensive and impractical to design the entire lighting system to provide such a high level of illumination; the best solution is therefore to mount a small luminaire directly on the machine or bench. Sometimes a highly concentrated 25- or 50-watt incandescent source is sufficient to provide the required light level.

When installing a supplementary system, the designer should remain within the luminance ratios recommended by the IES and given in Chap. 1. It takes a certain amount of time to adapt from one light level to another, so the differences in light level in a worker's field of view should not be too great. If they are, the worker's eyes will soon become fatigued and uncomfortable.

Improving the light levels can increase learning speed for a task and can also aid in inspection. Any random, nonrepetitive type of work will be benefited if the light level meets recommended levels.

However, for some tasks it is even more important to improve contrast. A worker who has sewn on pockets for years may do it almost by rote, so that the light level is not a critical factor. But perhaps the worker's eyes have aged, and it is more difficult to thread the needle of the sewing machine. In that case, developing contrast at the eye of the needle would be helpful.

In inspection lighting, the main goal is to establish the technique which works best for a particular problem. Once the technique is established, the light level can be adjusted upward to just below the point of glare.

Factors Involved in Vision

In order to design a supplementary lighting system, it is necessary to understand the factors involved in vision. This is particularly true in inspection lighting.

The four main factors are these:

1. Time
2. Size
3. Brightness
4. Contrast

TIME

It takes a certain amount of time for the eye to see an object. Less time is required if the eye must simply recognize an object's presence. More time is needed if the eye must focus on the object. Before the eye focuses, it must first find the object it is searching for. This is done by concentrating first on one spot and then progressing in a series of minute movements from one point to the next. When the object is found, the eye fixates.

If the object is moving, the eye must scan with it in order to produce an image. If the time required for this exceeds the length of time the object is in the field of view, it will be missed.

In inspection lighting, then, the inspector cannot be required to do more than his eyes are capable of doing. Since an inspector does not know when or where a defect will occur, the eyes are forced to carry on a continual search. This substantially reduces their ability to locate detail. There should be no attempt to perform any inspection which requires finding more than three objects a second, even under fairly ideal lighting conditions.

SIZE

An object's *visual size* is determined by the object's size and its distance from the observer. Most inspection problems involve locating the presence of a fault. If faults are too small to see quickly and comfortably, some form of magnifier can be used. If the object is far away, or hard to reach, closed-circuit television should be considered.

In the design of an inspection system, the actual size of the detail to be viewed is not very important, because similarly shaped objects having similar characteristics respond the same way to lighting regardless of their size. The lighting techniques used to bring out detail on surfaces are the same whether the detail is in a piece of cloth or on the side of a floodlighted building. These techniques are described later in this chapter.

BRIGHTNESS

The brightness of an object is determined by the amount of light which it reflects toward the eye. An object's brightness directly affects its visibility.

The amount of light which a surface reflects toward the eye is determined by the characteristics of the surface, the intensity of the light striking the surface, and the angle formed by the surface, the light source, and the eye.

Inspection lighting often involves techniques which are normally avoided in general lighting. For instance, some defects can be seen only if the brightness of the luminaire is reflected directly into the eyes of the inspector.

CONTRAST

Contrast is the visual difference between the object and the area surrounding it. It is the most important factor to consider when solving a difficult visibility problem. Visibility can be improved much more easily by improving contrast than by increasing task brightness. The success of the eventual solution may be determined by the ability to develop good contrast in the task.

The eye is attracted to things which have the highest contrast or the highest brightness. Therefore, it is necessary for objects that are being inspected to have these characteristics. The inspection techniques recommended in this chapter have all been developed with the aim of creating enough contrast between the object being sought and the surroundings for the inspector to perform the task quickly and effectively.

Organizing an Inspection Area

With these four factors in mind, the designer can more easily set up an efficient inspection area. To ensure good visibility, several initial steps should be taken.

1. Analyze the task. First determine exactly what the inspector is trying to see. The exact size, shape, texture, color, transmission properties, and reflectivity of the object and its background must be known. When they are, it is much easier to decide how the object should be lighted to emphasize the relevant detail. This is often an empirical task; the lighting designer often needs to view the task in its surroundings and try various lighting techniques.

2. Organize the inspection into the simplest possible form. Eliminate any unnecessary handling or manipulation of parts or movement by the inspector. Inspection alone demands intense concentration, and the inspector should not be asked to do other jobs as well.

3. If possible, key the inspection speed to the inspector rather than to the inspection line.

4. Eliminate any unnecessary obstructions from the visual scene. This is a particularly important point in conveyor operations, where the inspector should be able to see a defect the moment it enters the field of view. The defect should not appear and disappear, but should remain constantly visible to the inspector.

5. Train the personnel performing the visual function. This training generally will progress rapidly in the first week and become stable in about a month. A person can learn to identify rapidly several types of defects simultaneously if they are all equally visible. A trained observer can scan 5 to 10 times faster than an untrained observer.

Upgrading a lighting system can sometimes create problems. If an inspector has been trained under the original type of lighting, there is a good possibility that an initial reduction in speed may result even with improved lighting, since the inspector will have to relearn the job. In a short time, however, the inspector's speed will return to its former level or be improved.

6. If many parts are to be inspected, organize them so that they form a pattern and are oriented the same way. It is much easier to spot defects if the eye can scan along straight lines rather than in a random manner.

7. Consider installing some electronic or mechanical device to alert the inspector to the possible presence of a defect. Use of these devices could speed up the inspection process.

Position of the Light Source

The position of the light source for a specific inspection job depends on what needs to be seen. One placement is good for one task, and another for another.

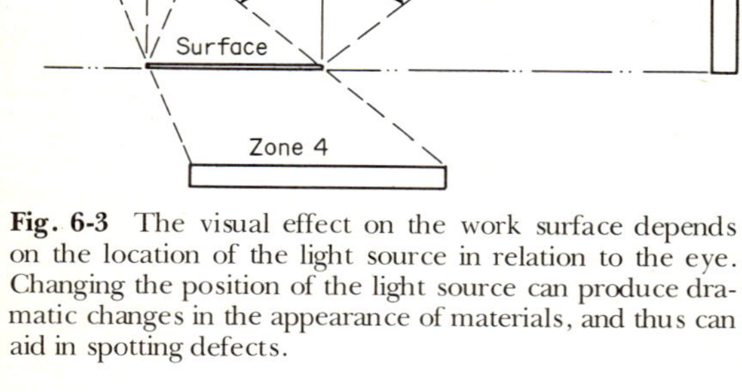

Fig. 6-3 The visual effect on the work surface depends on the location of the light source in relation to the eye. Changing the position of the light source can produce dramatic changes in the appearance of materials, and thus can aid in spotting defects.

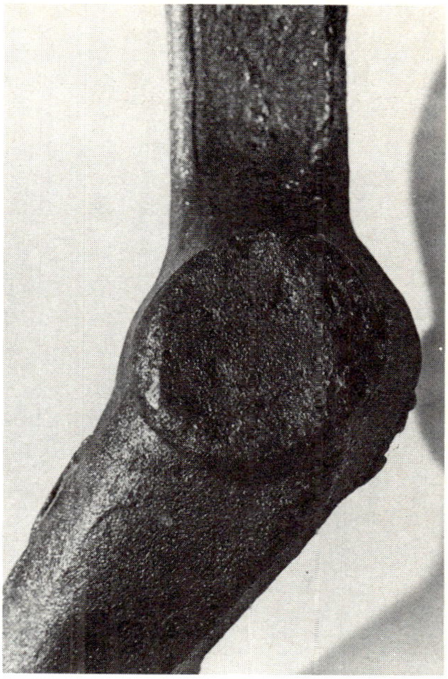

Fig. 6-4 This forging was lighted from sources in zones 1 and 2. This increased contrast by shadowing, so that the defect in the forging became apparent.

A geometric means for determining the position of a light source is shown in Fig. 6-3. The locations of the eye and the surface were arbitrarily chosen. The figure shows the surface in a horizontal plane. The boundaries of the zones are determined by the position of the surface in relation to the eye, and so a similar diagram with appropriate zones could be drawn for vertical or inclined planes as well.

Where curved surfaces are involved, the relationship becomes more complex. Where only a small area on a curved surface is involved, it can be analyzed by drawing the above relationship for the edges of the area and adjusting the boundaries of the zones accordingly.

Sometimes light sources can be mounted simultaneously in more than one zone or overlapping two zones for a particular inspection task. Often, though, placing a source in one zone will be sufficient.

ZONE 1

Light coming from zone 1, whether from diffuse, point, or linear sources, will reflect away from the eye. Only reflection produced by surface diffusion, refraction, or scattering of light will reach the eye and be visible.

Figure 6-4 shows a forging that was inspected with light coming from zones 1 and 2. The directional lighting created contrast with the background by shadowing, and so the fact that the forging was not filled out became obvious to the inspector.

Light sources are also placed in zone 1 if refraction is to be used to reveal defects. All transparent materials refract light, so large sections of plate glass can be inspected by using a point source of light placed in zone 1. Light is projected from the source through the glass and onto a screen. Any irregularity or fault in the glass will produce a shadow on the screen.

Color matching should be done by placing lighting in zone 1. If the source were placed in another zone, the lamp brightness reflected directly off the surface would mask the actual colors. The only exception to this rule is inspection of the color of polished metal surfaces. This should be done with reflected light coming from zone 2. Light sources in either case should be diffuse to eliminate problems associated with surface reflection of the light source.

Color differences are accentuated by a light source strong in the color which is most readily absorbed by the material. Slight color differences in red material, for example, are emphasized by sources strong in blue light and in blue material by light sources strong in red. Two different types of light sources, one in the blue end and one in the red end of the spectrum, are frequently used to check color matching. If the colors match under each source used separately, they will usually match under any source.

Fig. 6-5 Light from zone 2 was directed toward this ball joint. The reflected light emphasized the defective seam.

ZONE 2

Light sources mounted in zone 2 create contrast in the task by reflecting the light-source image pattern into the eye. This can be done with large-area, diffuse sources, light sources in a pattern, or a single directional incandescent, fluorescent, or HID source.

Diffuse lighting produces the most comfortable visibility, because there are no very bright or very dark areas. On surfaces whose color and general shape are being appraised, this type of lighting will produce the best results. It is most often used when specular metal surfaces are being inspected. Figure 6-5 shows how reflected light from a diffuse source emphasizes the defective seam in a ball joint.

Sources placed in zone 2 can be used for inspecting materials, such as black rubber, which absorb most of the light striking the surface. The only light which is seen is the light reflected directly off the surface because of cracks or defects.

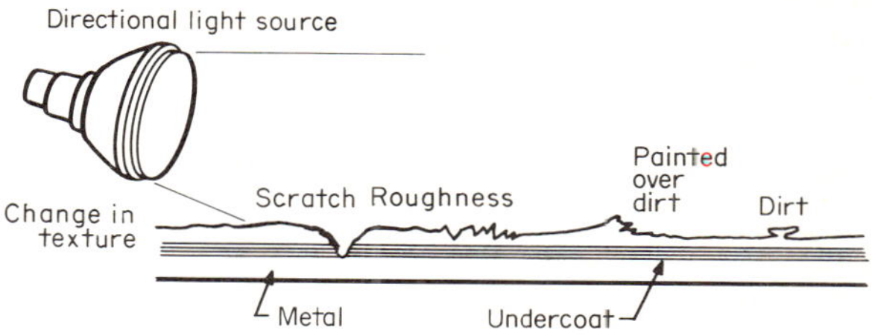

Fig. 6-6 A directional light source placed in zone 3 will highlight one side of a defect and place the other in partial shadow. This contrast makes the defects easy to see.

This method is also used in inspecting the surface of glassware, where the glassware is backed by a material having a very low reflectance, such as black velvet. The surface of glass reflects only about 4 percent of the light which strikes it, but this is enough to reveal imperfections in the surface if the background brightness is not too high.

When placed in zone 2, the light source should usually be positioned so that it is not visible in the inspector's line of sight. This greatly increases the visual comfort of the lighting system and improves task visibility through the absence of direct glare.

ZONE 3

Light coming from zone 3 will reflect below the level of the eye. This grazing lighting technique is illustrated in Fig. 6-6. This technique will exaggerate any surface detail on flat surfaces, such as the presence of dirt, unusual roughness, changes in texture, etc. The more concentrated the light source and the closer the light rays come to being parallel to the surface, the more these surface changes will be emphasized. For each of the defects one side will be highlighted and the other will be in partial shadow.

The size and angle of the light source can be changed to reveal almost any degree of surface roughness. Figure 6-7 shows how changing the size of the reflected image can change its effectiveness as an inspection tool. In the first case, a fluorescent luminaire was placed parallel to a cylinder's axis. This resulted in a narrow reflected image.

In the second case, the luminaire was placed perpendicular to the cylinder's axis, making a wide reflected image. The scratch in the surface is much more visible under the wider reflected image.

ZONE 4

Light sources are placed on zone 4 in order to project light through an object. This is useful when inspecting the insides of holes, such as those in metal parts. The first photograph in Fig. 6-8 shows a zinc casting placed on a light-colored bench top. It is difficult to see into the holes to inspect them. In the second photograph in Fig. 6-8, the same part is placed on a large-area, low-brightness lighting panel, with light sources behind it. Defects in holes now become apparent.

This technique is also used for inspecting translucent, transparent, or loosely woven material, such as glass or cloth. The light source is placed behind or below the material to be inspected, and changes in light transmission will indicate defects in the surface or variations in quality.

This technique of transillumination can be used for inspecting any material which will transmit light. In the textile industry it can be used to show fabric thickness variations and defects in weave.

Light-source intensity has to be adjusted to produce satisfactory bright-

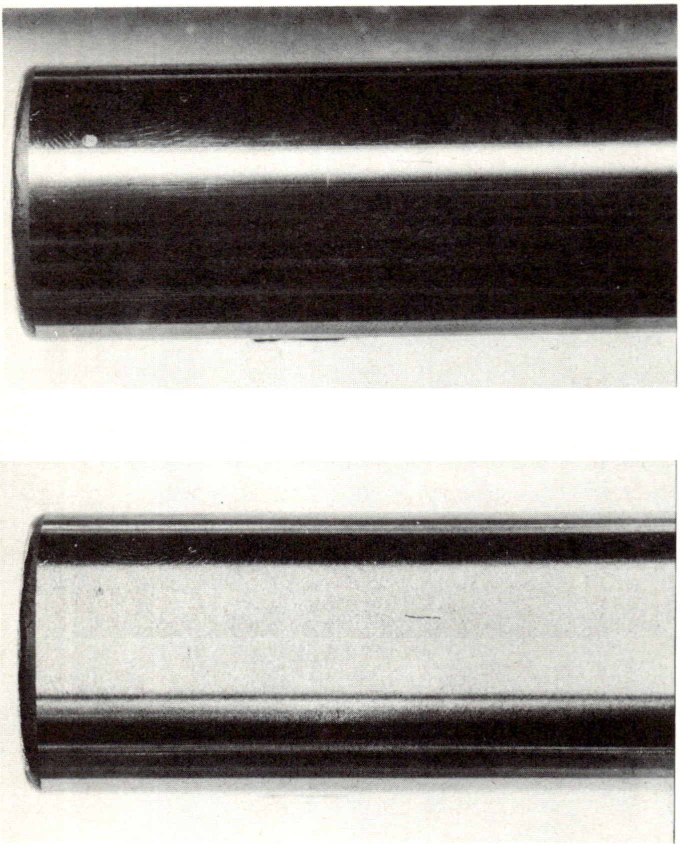

Fig. 6-7 Changing the size of a reflected image can increase its usefulness as an inspection tool. The scratch in the surface becomes more visible under the wider reflected image.

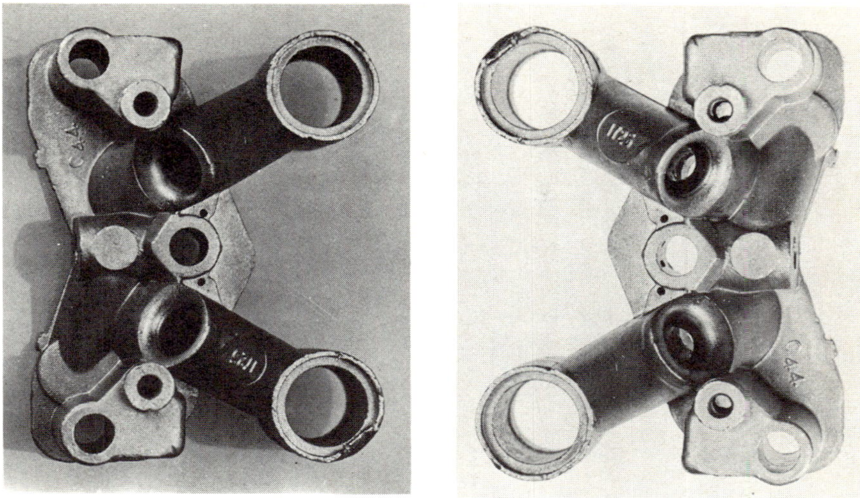

Fig. 6-8 In the first photograph, the zinc casting is being inspected on a light-colored bench top. It is difficult to spot any defects within the holes. In the second photograph, the same part is backlighted with sources placed in zone 4. Defects within the holes become apparent.

ness on the outside surface. The intensity on the outside surface should be between 2000 and 3000 lux (200 to 300 footcandles) in order for the system to be effective when the general lighting is turned on as well.

Bibliography

De Koker, Neil: "Supplementary Lighting Design Guide for Industrial Inspection Tasks," a thesis submitted to Chevrolet Central Office and General Motors Institute, Detroit, Michigan, July 1967.

De Koker, Neil, and John P. Frier: "Visibility of Specular and Semi-specular Tasks in Sheet Metal Surfaces," *Illuminating Engineering,* March, 1969, pp. 167–178.

Engel, F. L.: "Visual Conspicuity, Visual Search and Fixation Tendencies of the Eye," *Vision Research,* vol. 17, no. 1, 1977, pp. 95–108.

Frier, J. P.: "Revealing the Obscure," *Light Magazine,* vol. 33, no. 4, 1964, pp. 21–23.

Frier, John P.: "Difficult Industrial Seeing Tasks and How to Light Them," *Illuminating Engineering,* April 1967, pp. 283–288.

Hopkinson, R. G. and J. B. Collins: *The Ergonomics of Lighting,* Macdonald & Co., Publishers, Ltd., London, 1970, pp. 47–80.

"Lighting for Machining of Small Metal Parts," Illuminating Engineering Society of North America, New York, November 1949, p. 9.

7

Design Techniques for Long, Narrow Areas

A room cavity ratio (RCR) of 10 is the limit of coefficient of utilization (CU) tables used in the zonal-cavity method of lighting design. Some rooms have RCRs greater than 10; this is characteristic of long, narrow, deep areas where the mounting height is much greater than the width of the area. Warehouse aisles, loading docks, and some hazardous locations would fall into this category.

In such situations it is desirable to use calculation methods that are similar to those used for outdoor lighting in that only the direct contribution of the luminaires is considered. The additional illumination contributed by light reflected off the vertical walls or sides of stacks is considered to be a bonus and usually adds less than 10 percent to the illumination level. Design procedures for this type of area are included in this chapter.

Open storage areas where items are placed on the floor and there is no obstruction between these items and the luminaires should be treated in the same way as open industrial areas. The lighting design procedures outlined in Chaps. 4 and 5 should be followed for these areas and for locations where stacks are not so high as the width of the aisle.

This chapter includes methods of checking illumination levels at specific horizontal and vertical points. This technique is necessary in order to ensure that there is sufficient illumination at key locations. Luminaires are often randomly spaced in such areas as power plants, refineries, chemical plants, and rooms filled with pipes and trays. In areas such as hazardous locations it may be important to know the level of illumination on a control panel or on a valve or gauge that must be monitored.

Methods for checking point-by-point illumination levels are valid for areas of any size and shape. Techniques given in this chapter can be used to verify minimum illumination requirements or to check the level of illumination at a point in a room lighted with conventional design procedures.

Characteristics of Areas Covered in This Chapter

The sample problems in this chapter deal with warehouse lighting, but the design techniques apply to any other area where one of the following conditions is met: Either the room is square and has an RCR equal to or greater than 10; or the area is long, and the mounting height is equal to or greater than 2 times the width of the area. If the RCR is less than 10, the design techniques in Chaps. 4 and 5 should be used.

For design purposes, the reflectance off walls and stacks is considered to be 0. This is true of loading docks that are open on one or both sides, catwalks, and warehouse aisles with vertical racks that may or may not be full. If a rack is empty, the shelves form a horizontal baffle which will absorb any light that strikes them. In other cases, the reflectance value is not significant and can be ignored.

Levels of Illumination

Levels of illumination for particular situations can be read in Chap. 1. In a warehouse, if stacks are very high (that is, 12 to 15 meters or 40 to 50 feet), usually only enough light is needed for workers to identify portions of aisles and to load or unload material with a lift truck. In areas where material is placed into storage and retrieved by hand, however, stack height is usually about 3 meters (10 feet). Enough light must be provided to permit reading of labels and matching of invoices both in aisles and in stacks.

It is very difficult to produce vertical illumination levels over 100 lux at the base of warehouse stacks over 6.1 meters (20 feet) high. In this kind of warehouse, catalog numbers 3 centimeters (1 inch) high are recommended, and the racks and their contents should be clearly identified.

Luminaire Choice

Continuous rows of fluorescent luminaires have been used in the past for warehouse lighting. They can still be a good choice for stacks under 4.6 meters (15 feet) where small parts are picked out of bins. At higher stack heights the fluorescent systems will light only the upper portions of the stacks, since all fluorescent luminaires are wide-beam. This pattern of light distribution can be verified by calculating CUs, as explained in the next section.

High-intensity discharge (HID) systems will provide higher illumination levels on the aisles and lower portions of the vertical stacks while expending less energy. Before selecting a specific luminaire, the designer should determine where the critical tasks will be located. A different type of luminaire can be used if it is necessary only to light the aisles rather than to provide uniform lighting on the vertical stacks.

BEAM SPREADS

Wide-beam sources can be used to project light deep into vertical stacks or into boxcars or trucks, as shown in Fig. 7-1. High-pressure sodium (HPS) luminaires designed for low mounting heights were used for this application.

Fig. 7-1 Light projected into a boxcar.

Very narrow beam HID luminaires can direct light mainly onto the aisle floors. Other luminaires are somewhere in between, directing some light into the stacks and some onto the aisle floors.

Figure 7-2 shows a well-designed warehouse lighting system. The open HPS luminaires are on 9.8-meter (32-foot) centers and mounted at 13.7 meters (45 feet). They provide 215 lux (20 footcandles) on the aisle floor and 53.8 lux (5 footcandles) vertically on the bottom row of the stack.

Figure 7-3 shows the light distribution patterns for a number of 400-watt HPS luminaires with different beam spreads. The isoillumination distribution curves show where the light level will be 10.76 lux (1 footcandle) on the vertical surfaces of storage stacks. Each curve is marked with the spacing criterion (SC) for that particular luminaire. The setback is assumed to be 1.5 meters (5 feet).

SPACING CRITERION

As seen in Fig. 7-3, the greater the SC of the luminaire, the greater the amount of light delivered toward the tops of the stacks. The light level on the aisles and lower portions of the stacks will be higher with luminaires having lower SC numbers.

Selecting the proper luminaire is often a compromise. The more concentrating the distribution, the poorer the light distribution across the tops of the stacks. Newer HID luminaires are designed with asymmetrical beam patterns that throw more light up and down the aisles than sideways and provide better uniformity on the vertical stacks while still efficiently lighting the aisles and lower portions of the stacks.

SETBACK

If HPS sources are mounted on 6-meter (20-foot) centers at about the height of the stacks, approximately the top 1.5 meters (5 feet) of the stacks will not receive any light directly from the luminaires. If the luminaires are mounted slightly above the tops of the stacks, the dark spots will be eliminated. This higher mounting is usually required anyway to maintain adequate clearances. Generally it is best to mount the luminaires no

Fig. 7-2 A well-designed lighting system used for an indoor storage area.

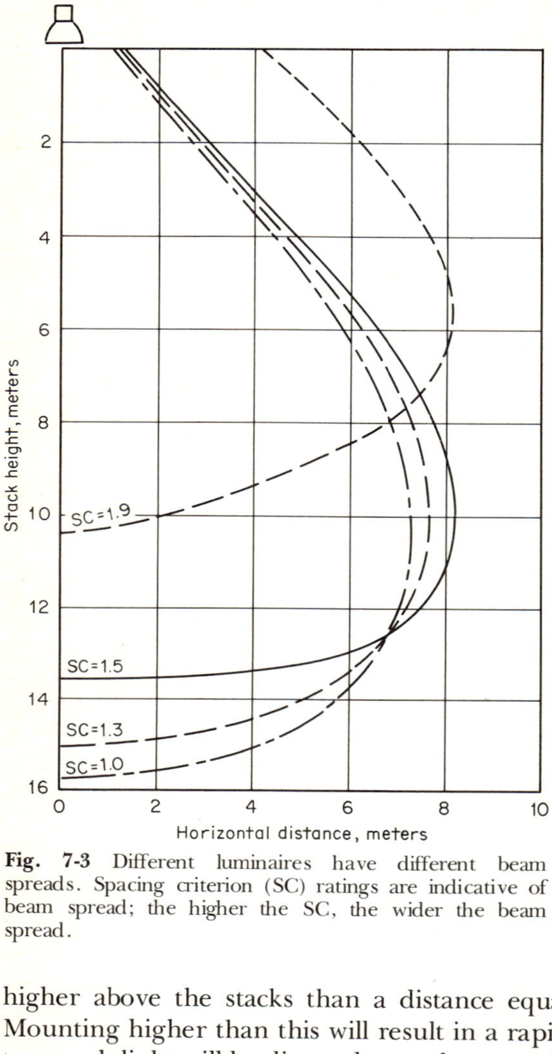

Fig. 7-3 Different luminaires have different beam spreads. Spacing criterion (SC) ratings are indicative of beam spread; the higher the SC, the wider the beam spread.

higher above the stacks than a distance equal to half the aisle width. Mounting higher than this will result in a rapid drop in utilization, since too much light will be directed onto the tops of stacks.

Determining Coefficients of Utilization and Spacing for Aisle Lighting

Table 7-1 can be used to determine the CU of a luminaire for that portion of the light which falls in an aisle between stacks. It is necessary to know the CU of the luminaire before the spacing can be calculated.

Table 7-1 Coefficient of utilization for illumination in an aisle

Column 1	Column 2	Column 3	Column 4	Column 5	Column 6	Column 7	Column 8	Column 9
			First Luminaire			Second Luminaire		
Zone	Degrees Vertical Angle	Multiplier	cd*	$2W$ / (LL × MH)*	Zone CU	cd	$2W$ / (LL × MH)	Zone CU
0–10	5	0.174	138	0.001	0.024	28,765	6.5×10^{-6}	0.033
10–20	15	0.169	153	0.001	0.026	29,503	6.5×10^{-6}	0.033
20–30	25	0.159	174	0.001	0.028	21,084	6.5×10^{-6}	0.022
30–40	35	0.144	177	0.001	0.025	15,173	6.5×10^{-6}	0.014
40–50	45	0.125	184	0.001	0.023	3,326	6.5×10^{-6}	0.003
50–60	55	0.102	160	0.001	0.016	98	6.5×10^{-6}	–
60–70	65	0.077	92	0.001	0.007	75	6.5×10^{-6}	–
70–80	75	0.051	42	0.001	0.002	64	6.5×10^{-6}	–
				Total CU	0.151			0.105

*Per 1000 lamp lumens.

The figures in the table were derived by dividing the aisle into zones on either side of the luminaire. Figure 7-4 shows how these zones were delineated. The first zone covers the area from 0 to 10 vertical degrees from the luminaire; the second is from 10 to 20 degrees; and so on.

In order for Table 7-1 to be valid, four conditions must be satisfied:

1. The level of illumination in the center of each zone is the average level in the zone, from one side of the aisle to the other and from end to end.

2. The luminous intensity or candlepower from the luminaire does not change significantly from 0 to 15 degrees.

3. The luminaire is mounted over the aisle.

4. Aisle width is less than one half the mounting height. (The chart would still be accurate if this were not the case, but it would probably be better to treat the aisle as a room.)

DERIVATION OF TABLE 7-1

The illumination level in the center of each zone can be calculated from the inverse-square law:

$$E = \frac{\text{cd} \cos^3 \theta}{\text{MH}^2}$$

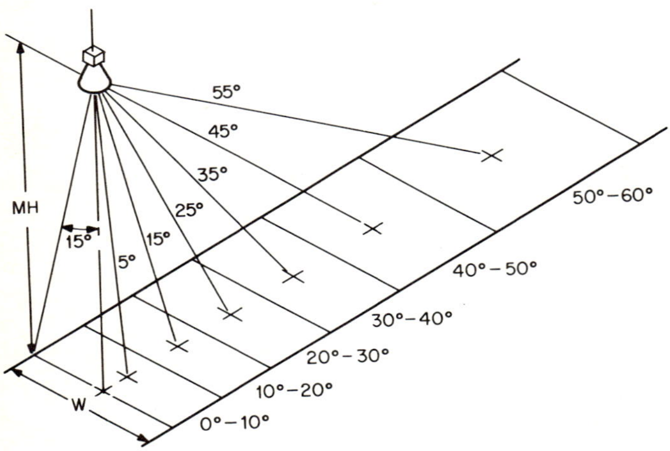

Fig. 7-4 Delineation of zones used in Table 7-1. The level of illumination in the center of each zone is considered to be the average level in the zone.

Since it is assumed that the average illumination in the entire zone is the same as the illumination in the center, the total lumens falling in the zone can be calculated by the following formula:

$$E = \frac{\text{lumens}}{\text{area}} \qquad \text{or} \qquad \text{Lumens} = E \times \text{area}$$

The area of the zone is the width of the aisle (W) times the length of each zone. The length (L) can be calculated, since the angle from the luminaire to the edge of each zone has already been selected:

$$L = \text{MH} (\tan b - \tan a)$$

where a = the angle to the near edge of each zone
b = the angle to the far edge of each zone

Thus, for the 10- to 20-degree zone in the example

$$L = \text{MH} (0.364 - 0.176) = \text{MH} \times 0.188$$

Combining the formulas

$$\frac{\text{cd} \cos^3 \theta}{\text{MH}^2} = \frac{\text{lumens}}{\text{MH} \times (\tan b - \tan a) \times W}$$

$$\text{Lumens} = \text{lamp lumens} \times \text{coefficient of utilization} = \text{LL} \times \text{CU}$$

$$\text{CU} = \frac{\text{cd} \times \cos^3 \theta \times (\tan b - \tan a) \times W}{\text{LL} \times \text{MH}}$$

This CU value would include the light on only one side of the luminaire; for both sides it would be twice as high. The value of $\cos^3 \theta \times (\tan b - \tan a)$ can be calculated for each zone and is tabulated in col. 3 of Table 7-1 as the zone multiplier. If the candela values are expressed per 1000 lumens, the lamp lumens (LL) should be 1000. If the photometric data is based on the actual lamp lumens, the lumen value shown on the photometric data should be used.

The CU for each zone is therefore

$$\text{CU} = \text{cd} \times \text{zone multiplier} \times \frac{2W}{\text{LL} \times \text{MH}}$$

The total CU is the sum of the CUs for all the zones.

This formula can be used for either HID or fluorescent luminaires and will work for warehouse aisles or any other long, narrow area where the mounting height is over twice the width.

SAMPLE PROBLEMS

Two sample problems are used throughout this chapter. The data for sample problem 1 are as follows:

Luminaire type: wide-beam 400-watt HPS
Lamp lumens: 50,000
Light-loss factor: 0.86
Illumination level: 538 lux (50 footcandles)
Mounting height: 6 meters (20 feet)
Aisle width: 3 meters (10 feet)
Task: handpick stock

The data for sample problem 2 are as follows:

Luminaire type: 400-watt HPS
Lamp lumens: 50,000
Light-loss factor: 0.81
Illumination level: 215.2 lux (20 footcandles)
Mounting height: 12 meters (40 feet)
Aisle width: 1.8 meters (6 feet)
Task: pallet-load stock

The luminaire in the first sample problem is listed as the First Luminaire in Table 7-1. The luminaire in the second problem is the Second Luminaire in Table 7-1.

EXPLANATION OF HEADINGS

The first column in Table 7-1 lists the zones into which the area is divided. The second lists the vertical angle from the luminaire to the centerpoint of each zone on the floor of a warehouse aisle.

The third column lists multipliers derived as explained above. It has been adjusted to allow for the fact that only a zone is being considered—not the entire surface subtended by the angle (except zone 1).

Columns 4, 5, and 6 are data for the luminaire in sample problem 1. Columns 7, 8, and 9 give similar data for the luminaire in sample problem 2. The width, mounting height, and lamp lumen data are used. Distribution data is taken from each luminaire's photometric data. Figure 7-5 gives the distribution data for sample problem 1.

Figures in cols. 4 and 5, or 7 and 8, are multiplied first together and then by the multiplier in col. 3 to find the zone CU as given in cols. 6 and 9. If the designer wished to find the CU for any other luminaire, aisle width, or mounting height, the appropriate figures could be substituted.

Distribution data

Vertical Angle	Candela	Zone	Zonal Lumens
		180–185	
180	1	175–180	0
175	1	170–175	0
170	0	165–170	0
165	0	160–165	0
160	0	155–160	0
155	0	150–155	0
150	0	145–150	0
145	0	140–145	0
140	0	135–140	0
135	0	130–135	0
130	2	125–130	1
125	5	120–125	3
120	7	115–120	4
115	8	110–115	4
110	8	105–110	4
105	8	100–105	4
100	7	95–100	4
95	9	90–95	5
90	11	85–90	7
85	18	80–85	12
80	28	75–80	18
75	42	70–75	26
70	63	65–70	38
65	92	60–65	53
60	128	55–60	67
55	160	50–55	75
50	179	45–50	74
45	184	40–45	68
40	181	35–40	59
35	177	30–35	52
30	176	25–30	44
25	174	20–25	36
20	166	15–20	26
15	153	10–15	17
10	142	5–10	10
5	138	0–5	3
0	137		

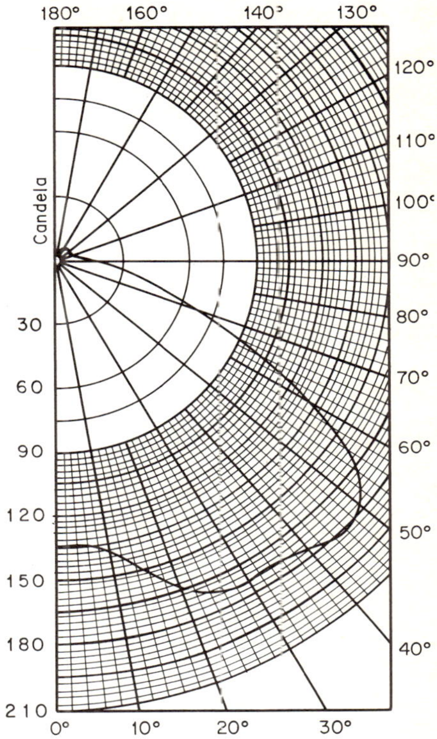

Tested at 25 feet with lamp rated at 1000 lumens. If data is desired for other lamp lumens multiply candela and lumen values given by ratio – different lamp rating divided by 1000 lumens.

Summary data

Zone	Lumens	Bare Lamp	Luminaire
90–180	29	3	4
0–90	686	69	96
0–60	533	53	75
0–30	137	14	19
0–180	715	71	100

Total efficiency = 71

Fig. 7-5 Distribution data and summary data for the luminaire used in sample problem 1.

SPACING

The figures in Table 7-1 can be used to arrive at correct spacing through the formula

$$E = \frac{\text{LL} \times \text{CU} \times \text{LLF}}{S \times W}$$

where E = level of illumination (lux or footcandles)
LL = initial lamp lumens
CU = coefficient of utilization in the aisle
LLF = light-loss factor (from manufacturer's data)
S = spacing (meters or feet)
W = width of the aisle (meters or feet)

For sample problem 1 a level of illumination of 538 lux (50 footcandles) is recommended. The lamp lumens for the 400-watt source to be used are 50,000. The total CU for this luminaire is 0.151, as read in col. 6 of the chart. The light-loss factor is 0.86. The spacing is unknown, and the aisle width is 3 meters (10 feet). The mounting height is 6 meters. The proposed luminaire choice has a wide beam spread, in order to project light into the stacks.

Substituting in the formula,

$$538 = \frac{50{,}000 \times 0.151 \times 0.86}{S \times 3}$$

$$538 = \frac{6493}{3\ S}$$

$$1614\ S = 6493$$

$$S = 4.0 \text{ meters}$$

This luminaire could be spaced at 4 meters (13.2 feet) and provide 538 lux on the aisle.

In sample problem 2 a narrower-beam 400-watt HPS luminaire is the proposed choice. In this case, stock will be loaded and unloaded with a forklift truck; the recommended level of illumination can therefore be 215.2 lux (20 footcandles). The initial lamp lumens are 50,000. The CU from Table 7-1 is 0.105, light-loss factor is 0.81, and aisle width is 1.8 meters.

Substituting in the formula,

$$215.2 = \frac{50{,}000 \times 0.105 \times 0.81}{S \times 1.8}$$

$$215.2 = \frac{4252.5}{1.8\ S}$$

$$387.36\ S = 4252.5$$

$$S = 11 \text{ meters}$$

This luminaire would provide the desired level of illumination on the aisle if it were spaced at 11 meters (36.3 feet).

Average Illumination on Vertical Stack

In warehouse lighting, it is often just as important to know the level of illumination on the stacks as it is to know the level on the aisle floors. Workers need to be able to see adequately when they handle materials stored in the stacks.

In order to determine the illumination on a stack, the designer must consult photometric data for the percentage of bare-lamp lumens which fall in the zone from 0 to 90 degrees. This information is given in the summary data. Figure 7-5 includes the summary data for the luminaire used in sample problem 1. It shows that 69 percent of the bare-lamp lumens fall in the area from 0 to 90 degrees. Subtracting the CU on the aisle floor, as given in Table 7-1, leaves the CU on the stacks. Therefore, the CU on the stacks is 0.69 − 0.151 = 0.54. This figure is for both stacks, and so it must be divided by 2 to yield the CU for one stack. Thus 0.54/2 = 0.27, or 27 percent of the bare-lamp lumens reach each stack.

If the luminaire is mounted at the top of the stack, then the average illumination level on the vertical stack can be determined by the formula

$$E = \frac{\text{LL} \times \text{CU} \times \text{LLF}}{S \times \text{MH}}$$

Substituting for the first example,

$$E = \frac{50{,}000 \times 0.27 \times 0.86}{4 \times 6}$$

$$E = \frac{11{,}610}{24}$$

$$E = 483.8 \text{ lux, or } 43.5 \text{ footcandles}$$

This is the average light level on each stack.

In the second example, the percentage of bare-lamp lumens from 0 to 90 degrees is 77. So the CU on each stack is (0.77 − 0.105)/2 = 0.33. The level of illumination on the stack is then

$$\frac{50{,}000 \times 0.33 \times 0.81}{11 \times 12} = \frac{13{,}365}{132} = 101 \text{ lux, or } 9 \text{ fc}$$

If the luminaire is mounted above the stack, the percentage of bare-lamp lumens from 0 to 60 degrees should be used.

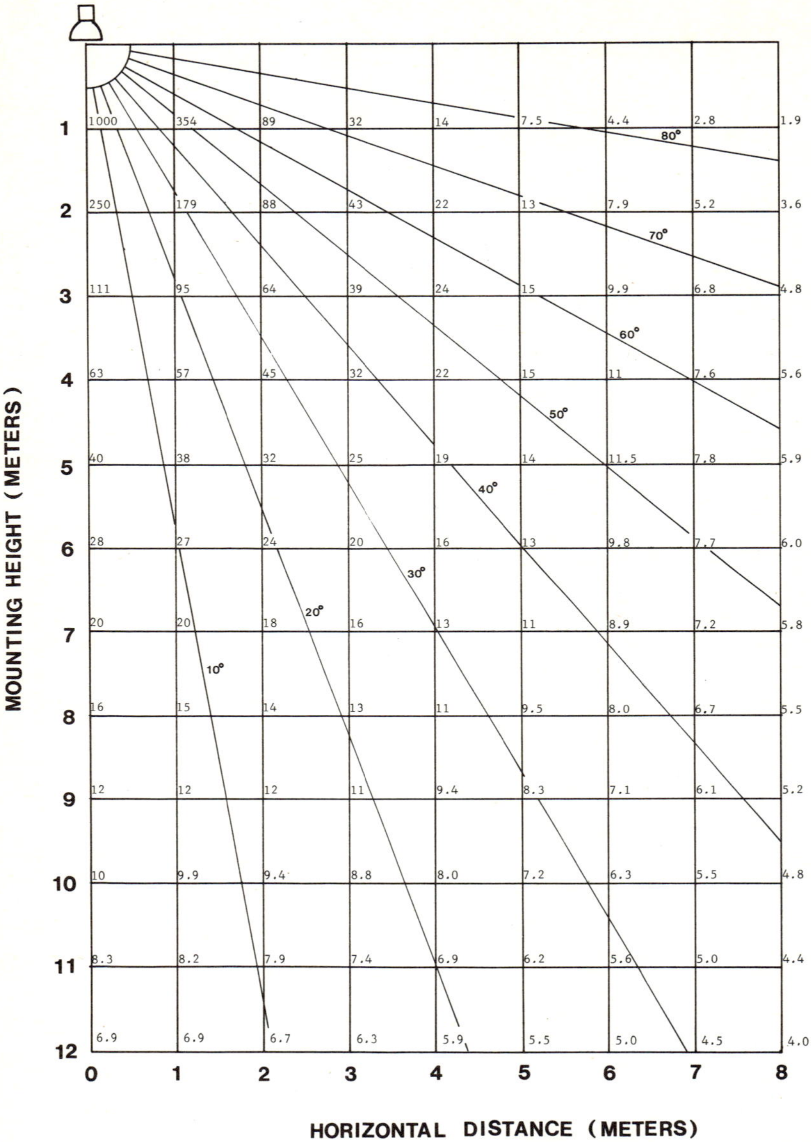

Fig. 7-6 Chart used to find the illumination level at any spot on a horizontal surface. Values are given in lux and meters but can be converted to footcandles and feet: 1 lux = 0.09 footcandle; 1 meter = 3.3 feet.

Checking a Point on the Horizontal Plane

Point-by-point calculation procedures are used to determine the expected level of illumination at a particular horizontal or vertical point. The inverse-square law and the cosine law have been combined in order to derive the two charts given below which enable the user to calculate horizontal and vertical illumination, respectively. The two laws combine to read

$$E = \frac{\text{cd} \cos^3 \theta}{D^2} \quad \text{or} \quad E = \frac{\text{cd} \cos^3 \theta}{\text{MH}^2}$$

where E = level of illumination at a point
cd = candelas directed toward that point
D and MH = the distance from the light source to the point
$\cos^3 \theta$ = the cosine of the angle of incidence of the light beam at the point being measured

Figure 7-6 enables the user to find the light level at a point on a horizontal surface. The two ordinates of Fig. 7-6 are the mounting height above the point and the horizontal distance away from it. Both are given in meters. If these values are in feet, they can be converted to meters by multiplying by 0.3048.

The numbers at each intersection of the chart are lux values per 1000 candelas at the specified point. These exactly correspond to footcandle values at the point, and can be converted to footcandles by multiplying by 0.09.

The diagonal lines are the angles from the luminaire to the point at which illumination is being measured.

In order to use the chart, the designer must know which luminaire is to be used and must have access to the distribution data for that luminaire. The mounting height above, and the horizontal distance away from, the point being measured must also be known.

Figure 7-7 is a layout of three luminaires in a warehouse aisle. They are spaced at 4 meters and mounted at a height of 6 meters. To find the light level at point P, follow this procedure:

1. For luminaire A, determine the luminaire mounting height above, and the horizontal distance away from, the point being measured. In this case, the mounting height is 6 meters and the horizontal distance is 2 meters.

2. Locate this point on the graph in Fig. 7-6 and find the illumination level per 1000 candelas at that point. In this case, the level is 24 lux.

3. From the graph, determine the vertical angle between this point and the luminaire. In this case, the angle is about 20 degrees.

4. From the distribution data for the luminaire being used, find the candelas at 20 degrees. The distribution data for the luminaire used in sample problem 1 is given in Fig. 7-5. The candelas at 20 degrees are approximately 166 per 1000 lumens.

5. From lamp data, find the initial lamp lumens for the luminaire. They are 50,000.

6. Determine the actual candelas at 20 degrees by the following formula:

$$\text{Candelas at } \theta \text{ per 1000 lumens} \times \frac{\text{LL}}{1000} = 166 \times \frac{50{,}000}{1000}$$
$$= 166 \times 50 = 8{,}300 \text{ candelas at 20 degrees}$$

7. Find the actual lux value at this point:

$$E \text{ from chart} \times \frac{\text{cd at } \theta}{1000} = \text{initial } E \text{ at } \theta$$

Since there are 24 lux per 1000 candelas at point P, as determined in step 2, then: 24 × 8300/1000 = 199.2 lux.

Since this figure is well below the recommended 538-lux value, proceed to step 8.

8. Add the contribution from the other two luminaires in Fig. 7-7. Since luminaire B is also 6 meters above and 2 meters horizontally from point P, its contribution is also 199.2 lux.

Luminaire C is 6 meters above and 6 meters horizontally from point P.

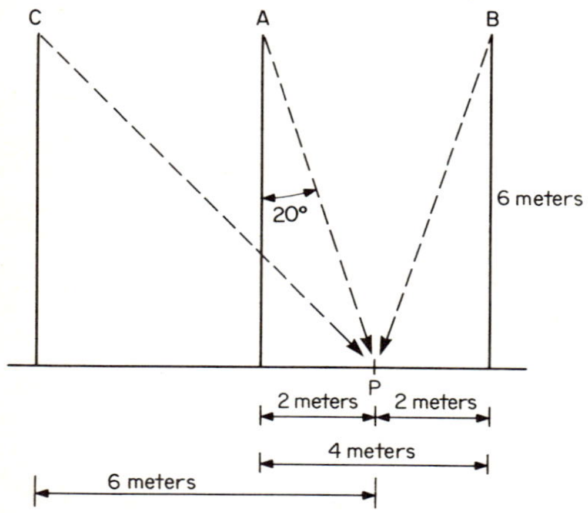

Fig. 7-7 The illumination level at point P is calculated by finding on Fig. 7-6 the individual contributions from each of the three luminaires A, B, and C.

Repeating steps 1 through 7 shows that luminaire C contributes 90 lux to point P.

The total contribution to point P from the three luminaires is 90 + 199 + 199 = 488 lux.

In addition, there will be some contribution from other luminaires in the row. The resulting illumination level is close enough to the recommended level of 538 lux to be satisfactory.

9. Multiply the level of illumination by the appropriate light-loss factor to find the maintained illumination level.

Checking a Point on the Vertical Plane

The vertical level of illumination at a point can be determined with the graph in Fig. 7-8. This graph can be used either in meters and lux or feet and footcandles. The values on the curved lines are either lux per 100 candelas or footcandles per 1000 candelas. The ordinates—vertical distance divided by setback and horizontal distance divided by setback—are ratios and are used with either meters or feet. One system must be used all the way through, since the values are not equivalent. The calculations below are consistently expressed in meters and lux.

The setback assumed for the chart is 3.16 meters, which is the square root of 10 meters. The angles on the flaring lines are the angles from the luminaire to the point being measured.

For example, assume that the layout is as shown in Fig. 7-9. Use Fig. 7-8 by following this procedure:

1. Determine the perpendicular setback from the A luminaire to the vertical plane being lighted. For this example, assume the setback is 1.5 meters.

2. Refer to Fig. 7-9 to determine the vertical and horizontal distances from A to the point P being measured. Divide each distance by the setback. In the example, point P is 2.7 meters down and 2.25 meters over; the ratios are therefore 2.7/1.5 = 1.8 (vertical distance) and 2.25/1.5 = 1.5 (horizontal distance).

3. Use these ratios to locate point P on the graph in Fig. 7-8. The illumination level is 0.6 lux per 100 candelas.

4. From the curved lines, determine the vertical angle between this point and the luminaire. In this example, the vertical angle is 45 degrees.

5. From the distribution data for the proposed luminaire choice, find the candelas at 45 degrees. The luminaire in the first example is still being used; its data is given in Fig. 7-5. This luminaire provides 134 candelas per 1000 lumens at 45 degrees.

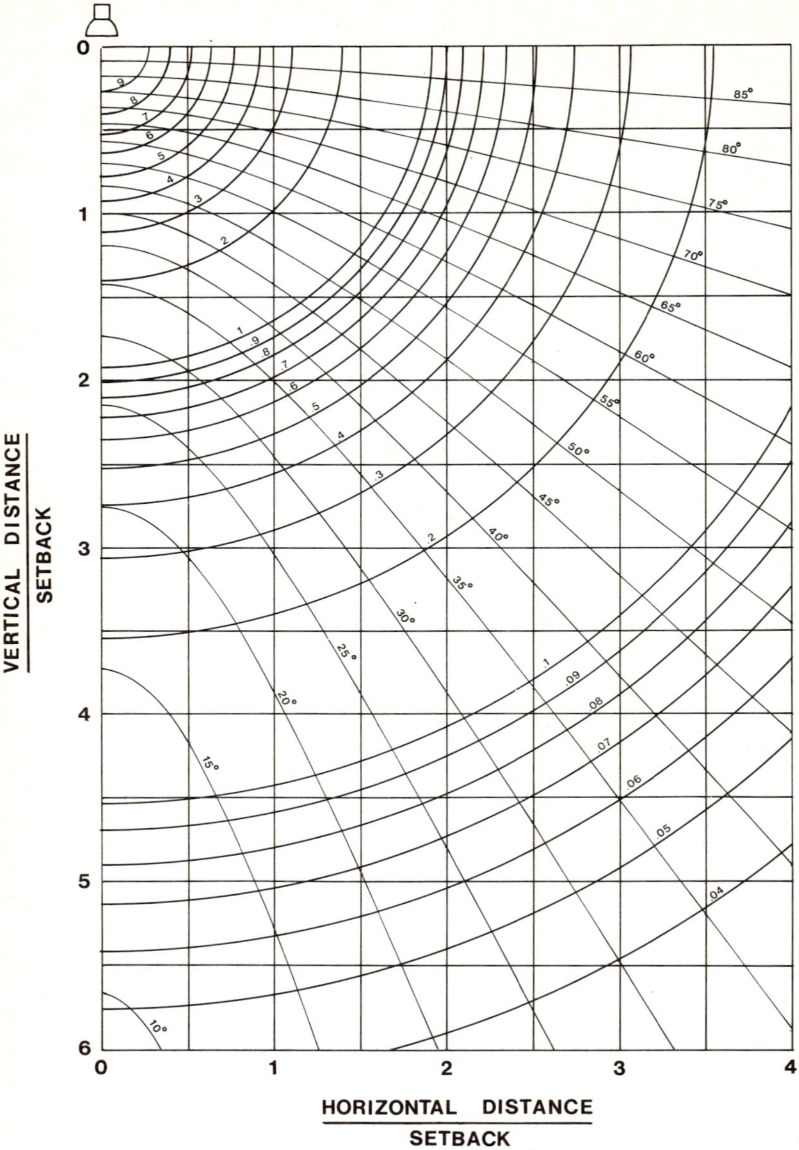

Fig. 7-8 Chart used to find the level of illumination at a vertical point. The setback assumed is 3.16 meters (the square root of 10), or 10 feet. Numbers on the curved lines are either lux per 100 candelas or footcandles per 1000 candelas. The chart must be used with one system of measurement all the way through.

6. From the lamp data, find the actual lumens and from this the actual candelas at point P. This luminaire is rated for 50,000 initial lamp lumens. Therefore $(50{,}000/1000) \times 184 = 9200$ candelas at point P.

7. Find the lux value at point P; $9200 \times 0.6/100 = 55.2$ lux. This is for a 1.5-meter setback. The graph is based on a setback of 3.16 meters. Find the appropriate multiplier in this way: $3.16^2/1.5^2 = 10/2.25 = 4.44$. The multiplier for any other setback is found in the same way.

Multiply the lux value at point P by the multiplier: $55.2 \times 4.44 = 245$ lux initial level at point P.

8. If desired, find and add the contribution from other luminaires.

9. Multiply by the appropriate light-loss factor to find the maintained value.

Hazardous-Location Lighting

Both Fig. 7-6 and Fig. 7-8 can be used to find horizontal or vertical levels of illumination in any kind of situation. They are particularly useful in hazardous-location lighting, where increased light levels may be needed

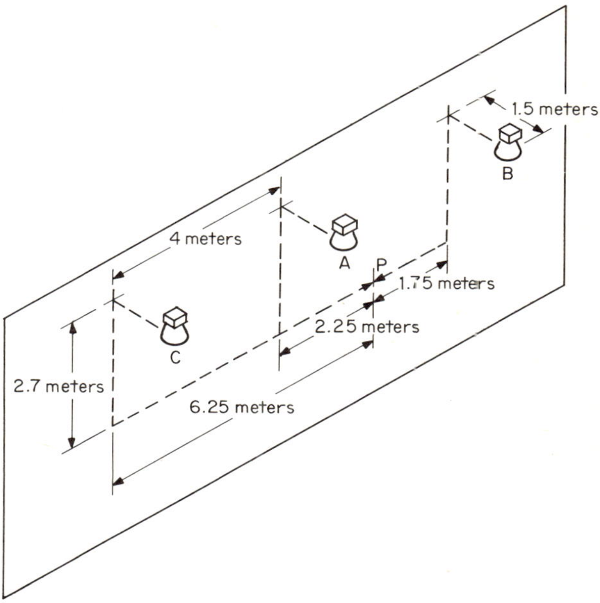

Fig. 7-9 Luminaire layout used in conjunction with Fig. 7-8.

only at specific points. Often, the general level of illumination can be quite low but a higher level is needed to read a dial, adjust a valve, or perform some other function. In such a case, photometric data from hazardous-location luminaires should be used in conjunction with the charts in order to find illumination levels at specific horizontal or vertical points.

8

Offices and Laboratories

There has probably been more study of office lighting than of all other types of lighting combined. The kind of visual task encountered in an office is typified by written or printed material at distances of 0.25 to 0.5 meters (10 to 20 inches). The worker in an office or laboratory is generally confined to a desk or worktable. The work is usually sedentary, requiring mental concentration rather than physical effort. As a consequence, distractions can be a greater problem than they are in a manufacturing area.

Basic lighting design procedures for offices and laboratories are the same as for any other area. However, there are differences because of the nature of the tasks performed and the generally lower ceiling heights. This chapter will discuss some of the differences, and the designer should incorporate this information into the design techniques explained in Chaps. 4 and 5 in order to design an office or laboratory lighting system.

Equivalent Sphere Illumination

Equivalent sphere illumination (ESI) is a concept that relates the visibility of a task to the illumination level. The lighting design method which considers only the illumination level does not take into account any loss of visibility that is due to veiling reflections.

Lighting systems that have equal illumination levels do not necessarily provide the same visibility. The difference is often due to a loss of task contrast because of veiling reflections. Veiling reflections are caused when the brightness of the luminaires and ceiling is reflected directly off the surface of the task. This reflected brightness is a function of the luminaire brightness as well as of the average illumination level. Luminaire placement is a significant factor. On such things as glossy paper, photographs, and material written in soft lead, veiling reflections can cause severe loss of task contrast and visibility. All office tasks are affected to some extent.

Equivalent sphere illumination uses a uniformly lighted hemisphere as a reference source of illumination. The uniform hemisphere produces the same surface and task brightness regardless of the orientation of the task which is placed beneath it.

Visibility measurements within the sphere are repeatable and can be used as a reference. The sphere measurements are compared with the visibility produced by the actual lighting system for the same task viewed from a certain direction at a particular point in a room. The visibility will be higher or lower than was produced in the sphere, depending on the effect of veiling reflections.

The ESI level for the task is that level of illumination that a photometric sphere would have to produce in order to make the target or task as visible in the sphere as it is in the real lighting environment. The illumination level in an office may be higher or lower than the ESI rating for the tasks performed. Reference or sphere lighting of 80 footcandles may produce the same visibility as 65 footcandles in an actual lighting system, or vice versa. If 65 footcandles produced by a lighting system provide the same task visibility as 80 footcandles of sphere lighting, the installation is said to provide 80 ESI even though 65 horizontal footcandles are produced by the lighting system. Equivalent sphere illumination levels will vary at different points in a room and at different viewing angles.

The Illuminating Engineering Society of North America (IES) now gives recommended levels of illumination for offices in terms of ESI. The measurement and calculation techniques for ESI levels are very complex. Computer programs are available which can calculate the ESI values for typical office layouts. A designer can consult equipment manufacturers or architectural design and consulting firms for assistance in designing a system to ESI levels. The design techniques explained in this chapter and in Chaps. 4 and 5 should be carefully followed to ensure satisfactory results.

The ESI concept applies only to tasks which lose contrast and visibility because of veiling reflections. In industrial laboratories or in manufacturing areas there may be no tasks that are affected in this way. Metal surfaces rely on direct reflected luminaire brightness to produce surface brightness. For certain type details in a metal surface, a uniformly lighted sphere can reduce the contrast of the task to 0. Considerably more research will be required before the ESI concept can be applied to all areas of lighting.

Luminaire Type and Placement

Veiling reflections can be reduced and ESI levels maximized by proper selection and location of luminaires in relation to the task. Research has shown that most office tasks are viewed at angles from 0 to 40 degrees, with 25 degrees being the most common viewing angle. If luminaires are placed in such a way that light reflects directly into a person's eyes, they are said to be in the *offending zone*. Luminaires placed in this zone will aggravate the problem of veiling reflections.

The offending zone is generally considered to be immediately in front of and above a person seated at a desk. Since the angle of incidence of the light equals the angle of reflection, the offending zone varies with the situation. If work is tilted rather than flat on the desk, the angle of reflection will be changed. The height of the viewer is also a factor.

In order to avoid placing luminaires in the offending zone, the designer must know the orientation of the work surface (or the plane of the task) and the location of the worker before designing an office lighting

Fig. 8-1 Luminaires in this office are placed to either side of the desk rather than in the offending zone. In this way, light will not be reflected directly into the eyes of the person working at the desk.

system. For instance, if office equipment is already in place or its placement is known, the designer can sit at the desk or desk location and place a flat mirror on the desk surface or at the level where the desk surface would be. Any portion of the ceiling that can be seen reflected in the mirror will fall in the offending zone, and luminaires should not be placed in that area. Placement of the luminaires in the office shown in Fig. 8-1 was done in this manner. Note that they are not immediately above and in front of the person seated at the desk; instead they bracket the work area.

If the lighting system in an office is already in place and there is a choice in desk placement, the process can be reversed. Mirrors can be placed on the desks, and the desks can be moved to locations where luminaires are not reflected in the mirrors. This is an easy and effective way to position desks and/or luminaires when an office is being occupied.

Because of the lower ceiling heights used in offices, the room cavity height is generally 2 to 3 meters (6 to 10 feet). To prevent shadows, luminaire spacing should be close to the mounting height. This dictates that the total lamp lumens per luminaire should be less than 20,000 (see Chaps. 4 and 5). The majority of office lighting systems use 2- or 4-lamp

1.2-meter (4-foot) fluorescent lamps. These have traditionally been 40-watt lamps, but recent improvements in lamp-phosphor and in ballast design have enabled the wattage of the lamps to be reduced with only a minor reduction in illumination. High-pressure sodium (HPS) systems of 100 and 150 watts can also be used for offices and laboratories. Their higher initial cost can be offset by power savings and reductions in air/conditioning requirements.

Small offices should be lighted with a minimum of two luminaires spaced to either side of the work location. In larger, open offices there is less opportunity to orient the luminaires in relation to the task location. In the latter case, if luminaires are located in the offending zone, a reduction in veiling reflections must be achieved by the light distribution of the luminaire. This is accomplished by using luminaires with asymmetrical distributions. The main beam is directed 90 degrees from the orientation of the desks in what is called a bat-wing distribution. In the luminaire this can be achieved by the reflector alone or by a combination of the reflector and refractor.

Open-bottom luminaires require special attention. The exposed brightness of the bare lamp will produce the highest veiling reflections, whether the source is fluorescent or high-intensity discharge (HID). The brightness of an HID lamp will also create problems with reflected glare if used with office equipment having shiny parts or glossy surfaces.

Tasks performed at drafting tables are some of the most difficult to light. There will always be some shadows produced, regardless of the lighting system. The shadows will be minimized by using large-area, low-brightness luminaires or a luminous ceiling. For very fine drafting a supplementary drafting-table light is a good solution.

Room Surfaces

Table 8-1 gives the reflectances of room surfaces recommended by the IES in the *American National Standard Practice for Office Lighting*. The reflectances of ceilings, walls, and floors can increase the utilization of light in an office or laboratory much more than they can in manufacturing areas. Room cavity ratios (RCRs) are much higher, since rooms are smaller, and light reflected off walls is closer to the work plane. In addition, room surfaces tend to stay cleaner in offices than in most industrial areas.

Colors should be chosen which will provide the recommended levels of reflectance. Brighter areas can be used for color accents, if they are restricted to about 10 percent of a person's total viewing area. Colors should be chosen under the light source which has been specified. If windows are present, the shielding media should have the same reflectance as that recommended for walls.

Matte finishes on desk tops or laboratory counters will reduce reflected glare, which is reflection in any shiny surface that can create distraction or discomfort within the visual field. Shiny surfaces should be avoided in business machines as well, if possible.

Vertical surfaces on desks, file cabinets, etc., should have light finishes in order to make them moderately bright, since they receive less direct light than horizontal surfaces.

Luminance ratios should not exceed those given in the chart in Chap. 1. Eyes should not have to adapt to excessive changes in luminance or they will tire and the worker will become uncomfortable.

Noise can be a distraction in an office environment. Therefore procedures given in Chap. 3 should be followed to determine the correct lighting system noise criterion for the luminaires.

The designer should try to make the office or laboratory environment a pleasant one, since this has a psychological effect on workers. They tend to perform more effectively in a pleasant, comfortable setting, particularly when tasks are sedentary and repetitive.

Furniture Mounting

In the past few years there have been a number of office furniture designs that combine work spaces and filing space, and room dividers. Frequently these systems include supplementary desk lighting. Other designs also include lighting systems that bounce light off the ceiling, making it unnecessary to install a general office-lighting system.

The supplementary lighting in these designs is beneficial, since it compensates for the ceiling lighting cut off by the partitions. The indirect lighting system, however, should be evaluated separately. The efficiency

Table 8-1 Recommended surface reflectances for offices

Surface	Reflectance Equivalent Range
Ceiling finishes*	80–90%
Walls	40–60%
Furniture	25–45%
Office machines and equipment	25–45%
Floors	20–40%

*Recommended reflectances are for finish only. Overall average reflectance of acoustic materials may be somewhat lower.

SOURCE: Reprinted with permission from the *IES Lighting Handbook,* 5th ed., Illuminating Engineering Society of North America, New York, 1972, p. 11-5.

of an indirect system is lower than that of a direct system. This can be overcome somewhat by using lamps with higher efficacy. In general, the combination of a direct ceiling system with a supplementary system will be more efficient than an indirect system alone.

Integrated Ceiling Systems

In offices or laboratories with suspended ceilings it is possible to combine the lighting system with the heating and ventilating system. In larger offices it is also possible to use the heat generated by the lighting system to provide a portion of the total heating. In warm weather the lighting heat can be removed.

Bibliography

"American National Standard Practice for Office Lighting," *Journal of the Illuminating Engineering Society,* October 1973, pp. 3–40.

Design Practice Committee of the Illuminating Engineering Society of North America, "Recommended Practice for the Specification of an ESI Rating in Interior Spaces when Specific Task Locations are Unknown," *Journal of the Illuminating Engineering Society,* January 1977, pp. 111–123.

DiLaura, D. L.: "On the Computation of Equivalent Sphere Illumination," *Journal of the Illuminating Engineering Society,* January 1975, pp. 129–149.

Henderson, R. L., J. F. McNelis, and H. G. Williams: "A Survey of Important Visual Tasks in Offices," *Lighting Design And Application,* January 1975, pp. 18–25.

LeVere, Richard C.: "ESI—A Refresher in the Basics," *Lighting Design and Application,* July 1976, pp. 39–43.

Lewin, Ian: "An ESI Study for Different Tasks," *Journal of the Illuminating Engineering Society,* October 1976, pp. 44–51.

Murphy, Amy: "Trends in Furniture-integrated Lighting Design," *Lighting Design and Application,* February 1977, pp. 26–33.

"Offices and Schools," *IES Lighting Handbook,* 5th ed., sec. 11, Illuminating Engineering Society of North America, New York, 1972.

9

Floodlight Photometric Performance

The design of a floodlighting system is similar to many engineering system-design procedures. First the performance of individual parts of the system is analyzed. Then this information is used to determine a final system design.

Included in this chapter are procedures that can be used to evaluate the performance of individual floodlights. When this data is thoroughly understood, the designer can proceed to the following chapter, Chap. 10, which shows how to use this information to arrive at a system design. If the suggested design procedures are followed, there will be less chance of overdesigning or underdesigning the system. Overdesigning increases costs and wastes power, and underdesigning can result in reduced visibility.

There are variations among floodlights, however—even among those produced by the same manufacturer. There can be as much as a 20 percent variation between calculated levels of illumination and those actually measured, even when design procedures have been carefully followed.

The following recommendations are intended as guidelines only. No individual floodlight will meet all design criteria. Usually, compromises must be made or more than one type of floodlight must be used in the design of a system. Floodlights are almost always used outdoors, though they can be used indoors if the ballast is rated for indoor use.

Types of Systems

There are two types of high-intensity discharge luminaires used for outdoor lighting: floodlights, which can be aimed, and luminaires with fixed aiming, such as roadway or high-mast types. Floodlighting will be covered in this chapter and the following one. High-mast and roadway lighting systems are analyzed in Chaps. 11 and 12.

Fixed lighting systems are more efficient than floodlighting systems, but less flexible. Roadway luminaires work best on long, relatively narrow areas, such as roads or fences, where the width of the lighted area is less than 2 times the mounting height of the luminaire. For greater efficiency, both roadway and high-mast systems should be located inside the area to be lighted. In general, floodlights are used where the poles are located on the perimeter of the lighted area, or where illumination must be provided at a distance of more than 2 times the mounting height from the base of the floodlight location.

No one system must be used exclusively. Floodlights can be mounted on the same pole with high-mast and/or roadway luminaires. This is often done when lighting large or irregularly shaped areas. In order to design the system, the designer can compute the contribution of each type of luminaire separately and add the results.

Floodlights are available in a variety of beam shapes, using incandescent and high-intensity discharge (HID) lamps in all of the common wattages and lumen ratings. So many different models are available that it is essential to understand how to use manufacturer's published photometric data. In this way the characteristics of various floodlights can be compared and matched with the needs of a particular situation.

Presentation of Photometric Data

Figures 9-1 and 9-2 show photometric data for two typical floodlights: the first is symmetrical, with a circular beam pattern, and the second is asymmetrical, with a beam pattern that is wider in one direction than in the other. The format of the data follows guidelines published by the Illuminating Engineering Society of North America (IES). This general method of presentation is used by independent testing laboratories and by most manufacturers.

The current method of presenting data shows the photometric performance of the luminaires on the basis of a constant number of lamp lumens, regardless of the lamp which is actually used in the luminaire. This is known as the *relative method.* This approach is somewhat more complicated than the *direct method,* which uses the actual lamp lumens. The direct method was generally used before 1971 and is still sometimes used. Some photometric data gives a summary based on both approaches.

The relative method has two main advantages which ensure greater accuracy:

1. In order to use the photometric data, the designer must obtain the latest published lamp lumen rating from the lamp manufacturer. Since these values are frequently changed, the relative method ensures that the most recent value will be used.

2. Since all curves are based on the same lamp lumens, it is easy to make comparisons between floodlights.

Figures 9-1 and 9-2 are based on a constant of 1000 lamp lumens.

Candela and Lumen Distribution

On the upper left side of the photometric data in Figs. 9-1 and 9-2 is the tabulation of the photometric performance of the luminaire. This information is given for a floodlight located above the 0- by 0-degree point on the right side of the table. The floodlight is aimed directly down at this 0-degree point. Since a floodlight beam is usually symmetrical left and right,

LIGHTING SYSTEMS DEPARTMENT
HENDERSONVILLE, N. C.

GENERAL ELECTRIC

PHOTOMETRIC DATA PER 1000 LAMP LUMENS

CANDELA AND LUMEN DISTRIBUTION — AVG. LEFT AND RIGHT SIDES

LUMINOUS INTENSITY IN CANDELAS AT CENTERS OF ZONES (UPPER NOS.)
LUMINOUS FLUX IN LUMENS WITHIN ZONES (LOWER NOS.)

HORIZONTAL ANGLES — DEGREES

% LAMP LUMENS	LUMENS — SUMMATION OF ONE SIDE ONLY	85.0	75.0	65.0	55.0	45.0	35.0	25.0	15.0	5.0	0.0	VERTICAL ANGLES — DEGREES
32.0	0.00	0.00 / 0.000	0.00 / 0.000	0.00 / 0.000	0.00 / 0.000	0.00 / 0.000	0.00 / 0.000	0.00 / 0.000	0.00 / 0.000	0.00 / 0.000	0.59	85.0 (UPPER BEAM)
32.0	0.25	0.00 / 0.000	0.00 / 0.000	0.00 / 0.000	0.29 / 0.005	0.88 / 0.019	1.18 / 0.029	1.77 / 0.049	2.36 / 0.069	2.65 / 0.080	2.95	75.0
31.9	2.09	0.00 / 0.000	0.00 / 0.000	0.88 / 0.011	2.36 / 0.041	6.48 / 0.139	11.49 / 0.286	15.61 / 0.430	19.14 / 0.563	20.32 / 0.616	21.21	65.0
31.5	5.29	0.00 / 0.000	0.29 / 0.002	2.65 / 0.034	10.01 / 0.175	19.14 / 0.412	28.86 / 0.719	38.58 / 1.064	46.24 / 1.359	50.36 / 1.526	51.84	55.
30.5	11.07	0.00 / 0.000	0.59 / 0.005	7.95 / 0.102	20.91 / 0.365	35.93 / 0.773	54.19 / 1.351	78.05 / 2.152	100.73 / 2.960	111.04 / 3.365	113.69	45.
28.3	18.85	0.00 / 0.000	1.18 / 0.009	11.78 / 0.151	31.22 / 0.545	55.08 / 1.185	93.66 / 2.334	144.02 / 3.971	172.59 / 5.072	184.37 / 5.588	185.55	35.
24.5	28.44	0.00 / 0.000	2.95 / 0.023	17.67 / 0.227	40.64 / 0.709	81.58 / 1.755	144.61 / 3.604	204.69 / 5.644	262.13 / 7.703	289.52 / 8.774	293.35	25.
18.8	39.68	0.00 / 0.000	4.12 / 0.032	20.91 / 0.269	48.89 / 0.853	103.67 / 2.230	177.89 / 4.433	265.66 / 7.325	365.80 / 10.749	455.04 / 13.791	468.88	15.
10.9	54.29	0.00 / 0.000	4.71 / 0.037	23.27 / 0.299	55.08 / 0.961	119.58 / 2.572	198.80 / 4.954	306.89 / 8.462	472.42 / 13.882	762.82 / 23.119	871.79	5.
		0.00	4.71	24.45	55.96	120.46	203.81	313.37	488.32	844.99	1103.87	0.
10.5	52.40	0.00 / 0.000	5.60 / 0.044	24,45 / 0.314	55.96 / 0.976	119.28 / 2.566	201.75 / 5.028	307.48 / 8.478	462.11 / 13.579	706.56 / 21.414	792.86	5. (LOWER BEAM)
18.2	38.65	0.00 / 0.000	4.71 / 0.037	22.38 / 0.288	49.48 / 0.863	102.49 / 2.205	177.30 / 4.419	260.95 / 7.195	354.31 / 10.412	436.48 / 13.229	445.32	15.
23.8	28.02	0.00 / 0.000	4.12 / 0.032	18.85 / 0.242	41.82 / 0.730	80.99 / 1.742	142.84 / 3.560	200.28 / 5.522	255.94 / 7.521	285.98 / 8.667	286.87	25.
27.5	18.64	0.00 / 0.000	3.53 / 0.028	14.43 / 0.186	31.51 / 0.550	55.08 / 1.185	94.25 / 2.349	140.49 / 3.874	171.71 / 5.046	179.07 / 5.427	180.84	35
29.7	10.79	0.00 / 0.000	2.36 / 0.019	8.84 / 0.114	21.21 / 0.370	36.52 / 0.786	54.78 / 1.365	77.17 / 2.128	99.84 / 2.934	101.61 / 3.080	101.32	45.
30.8	5.24	0.00 / 0.000	1.18 / 0.009	4.12 / 0.053	10.31 / 0.180	20.03 / 0.431	29.45 / 0.734	38.88 / 1.072	46.83 / 1.376	45.65 / 1.384	47.12	55.0
31.2	2.15	0.00 / 0.000	0.29 / 0.002	1.47 / 0.019	3.24 / 0.057	7.36 / 0.158	12.08 / 0.301	16.20 / 0.447	19.44 / 0.571	19.73 / 0.598	21.21	65.0
31.2	0.25	0.00 / 0.000	0.00 / 0.000	0.00 / 0.000	0.29 / 0.005	0.88 / 0.019	1.18 / 0.029	1.77 / 0.049	2.36 / 0.069	2.65 / 0.080	2.95	75.0
31.2	0.00	0.00 / 0.000	0.00 / 0.000	0.00 / 0.000	0.00 / 0.000	0.00 / 0.000	0.00 / 0.000	0.00 / 0.000	0.00 / 0.000	0.00 / 0.000	0.59	85.0
	316.11	0.00	0.28	2.31	7.39	18.18	35.50	57.86	83.87	110.74		
LUMENS — SUMMATION OF VERTICAL ZONES												
% LAMP LUMENS - LEFT OR RIGHT		31.6	31.6	31.6	31.4	30.6	28.8	25.3	19.5	11.1		

TEST MADE IN ACCORDANCE WITH IES APPROVED METHOD FOR PHOTOMETRIC TESTING OF FLOODLIGHTS

	IES (1952) BASED ON 10% AVG. MAXIMUM CP.	IES (1971) BASED ON 10% MAXIMUM CP.	
BEAM FLUX	570	518	LUMENS
BEAM EFFICIENCY	57	52	%
BEAM SPREAD	110°H X107°V	92°H X89°V	
NEMA TYPE	6 X 6	5 X 5	

NUMBER — REV. NO.

VERTICAL THROUGH 0 °HOR. CANDELA TRACES HORIZONTAL THROUGH 0 °VERT.

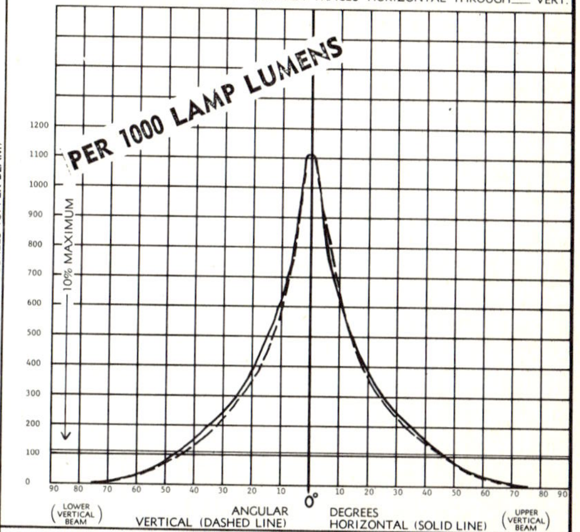

LUMINOUS INTENSITY (CANDELAS, CP.) 572.67 AVG. MAX. 1103.87 ABS. MAX.

VALUES OF CANDELA AND LUMENS ARE BASED UPON A LAMP OPERATED AT 1000 LUMENS.
IF THE DATA IS DESIRED FOR A DIFFERENT LAMP LUMEN RATING MULTIPLY ALL CANDELA AND LUMEN VALUES BY THE RATIO: DIFFERENT LAMP LUMEN RATING ÷ 1000

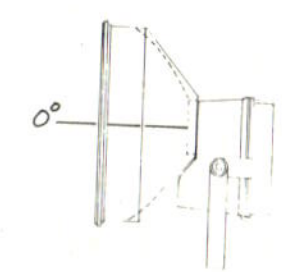

FLOODLIGHT DESCRIPTION

GENERAL ELECTRIC POWR-SPOT LUMINAIRE
REFLECTOR: 35-231870-03
FLAT CLEAR DOOR GLASS

NEMA TYPE: 5 X 5 (6 X 6)

LAMP: 1000 WATT HIGH PRESSURE SODIUM
G.E. NO. LU1000/BU (LUCALOX)

NUMBER — REV. NO.

Fig. 9-1 Photometric data for a 1000-watt high-pressure sodium (HPS) symmetrical floodlight.

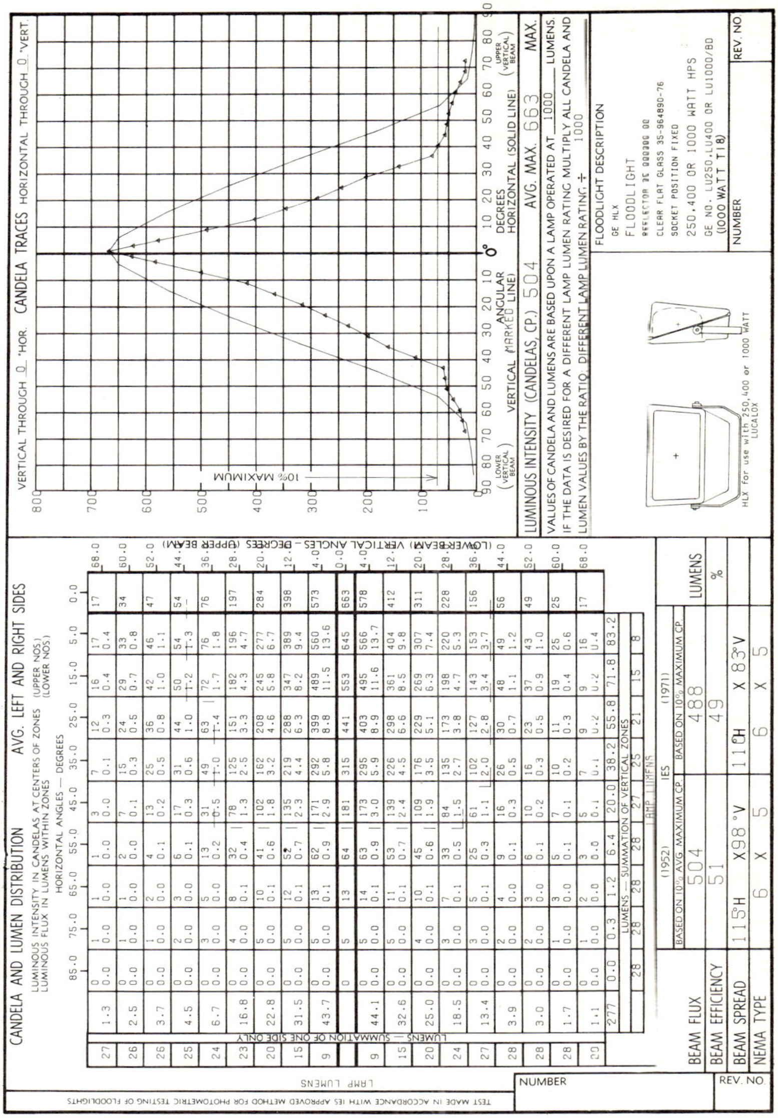

CANDELA AND LUMEN DISTRIBUTION — AVG. LEFT AND RIGHT SIDES

LUMINOUS INTENSITY IN CANDELAS AT CENTERS OF ZONES (UPPER NOS.)
LUMINOUS FLUX IN LUMENS WITHIN ZONES (LOWER NOS.)

HORIZONTAL ANGLES — DEGREES

LAMP LUMENS	LUMENS — SUMMATION OF ONE SIDE ONLY	85.0	75.0	65.0	55.0	45.0	35.0	25.0	15.0	5.0	0.0	VERTICAL ANGLES — DEGREES
27	1.3	0 / 0.0	1 / 0.0	1 / 0.0	1 / 0.0	3 / 0.0	7 / 0.1	12 / 0.3	16 / 0.4	17 / 0.4	17	68.0 (UPPER BEAM)
26	2.5	0 / 0.0	1 / 0.0	1 / 0.0	2 / 0.0	7 / 0.1	15 / 0.3	24 / 0.5	29 / 0.7	33 / 0.8	34	60.0
26	3.7	0 / 0.0	1 / 0.0	2 / 0.0	4 / 0.1	13 / 0.2	25 / 0.5	36 / 0.8	42 / 1.0	46 / 1.1	47	52.0
25	4.5	0 / 0.0	2 / 0.0	3 / 0.0	6 / 0.1	17 / 0.3	31 / 0.6	44 / 1.0	50 / 1.2	54 / 1.3	54	44.0
24	6.7	0 / 0.0	3 / 0.0	5 / 0.0	13 / 0.2	31 / 0.5	49 / 1.0	63 / 1.4	72 / 1.7	76 / 1.8	76	36.0
23	16.8	0 / 0.0	4 / 0.0	8 / 0.1	32 / 0.4	78 / 1.3	125 / 2.5	151 / 3.3	182 / 4.3	196 / 4.7	197	28.0
20	22.8	0 / 0.0	5 / 0.0	10 / 0.1	41 / 0.6	102 / 1.8	162 / 3.2	208 / 4.6	245 / 5.8	277 / 6.7	284	20.0
15	31.5	0 / 0.0	5 / 0.0	12 / 0.1	52 / 0.7	135 / 2.3	219 / 4.4	288 / 6.3	347 / 8.2	389 / 9.4	398	12.0
9	43.7	0 / 0.0	5 / 0.0	13 / 0.1	62 / 0.9	171 / 2.9	292 / 5.8	399 / 8.8	489 / 11.5	560 / 13.6	573	4.0
		0	5	13	64	181	315	441	553	645	663	0.0
9	44.1	0 / 0.0	5 / 0.0	14 / 0.1	63 / 0.9	173 / 3.0	295 / 5.9	403 / 8.9	495 / 11.6	566 / 13.7	578	4.0 (LOWER BEAM)
15	32.6	0 / 0.0	5 / 0.0	11 / 0.1	53 / 0.7	139 / 2.4	226 / 4.5	298 / 6.6	361 / 8.5	404 / 9.8	412	12.0
20	25.0	0 / 0.0	4 / 0.0	10 / 0.1	45 / 0.6	109 / 1.9	176 / 3.5	229 / 5.1	269 / 6.3	307 / 7.4	311	20.0
24	18.5	0 / 0.0	3 / 0.0	7 / 0.1	33 / 0.5	84 / 1.5	135 / 2.7	173 / 3.8	198 / 4.7	220 / 5.3	228	28.0
27	13.4	0 / 0.0	3 / 0.0	5 / 0.1	25 / 0.3	61 / 1.1	102 / 2.0	127 / 2.8	143 / 3.4	153 / 3.7	156	36.0
28	3.9	0 / 0.0	2 / 0.0	4 / 0.0	9 / 0.1	16 / 0.3	26 / 0.5	30 / 0.7	48 / 1.1	49 / 1.2	56	44.0
28	3.0	0 / 0.0	2 / 0.0	3 / 0.0	6 / 0.1	10 / 0.2	16 / 0.3	23 / 0.5	37 / 0.9	43 / 1.0	49	52.0
28	1.7	0 / 0.0	1 / 0.0	3 / 0.0	5 / 0.1	7 / 0.1	10 / 0.2	11 / 0.3	19 / 0.4	25 / 0.6	25	60.0
20	1.1	0 / 0.0	1 / 0.0	2 / 0.0	3 / 0.0	5 / 0.1	7 / 0.1	9 / 0.2	9 / 0.2	16 / 0.4	17	68.0
	277	0.0	0.3	1.2	6.4	20.0	38.2	55.8	71.8	83.2		
LUMENS — SUMMATION OF VERTICAL ZONES												
LAMP LUMENS		28	28	28	28	27	25	21	15	8		

	(1952) BASED ON 10% AVG. MAXIMUM CP.	IES (1971) BASED ON 10% MAXIMUM CP.	
BEAM FLUX	504	488	LUMENS
BEAM EFFICIENCY	51	49	%
BEAM SPREAD	115°H X98°V	110°H X 83°V	
NEMA TYPE	6 X 5	6 X 5	

LUMINOUS INTENSITY (CANDELAS, CP.) 504 AVG. MAX. 663 MAX.

VALUES OF CANDELA AND LUMENS ARE BASED UPON A LAMP OPERATED AT 1000 LUMENS. IF THE DATA IS DESIRED FOR A DIFFERENT LAMP LUMEN RATING MULTIPLY ALL CANDELA AND LUMEN VALUES BY THE RATIO: DIFFERENT LAMP LUMEN RATING ÷ 1000

FLOODLIGHT DESCRIPTION
GE HLX
FLOODLIGHT
REFLECTOR 35 000306 00
CLEAR FLAT GLASS 35-964890-76
SOCKET POSITION FIXED
250.400 OR 1000 WATT HPS
GE NO. LU250.LU400 OR LU1000/BD
(1000 WATT T18)

NUMBER | REV. NO.

TEST MADE IN ACCORDANCE WITH IES APPROVED METHOD FOR PHOTOMETRIC TESTING OF FLOODLIGHTS

Fig. 9-2 Photometric data for a 250-, 400-, or 1000-watt HPS asymmetrical floodlight.

only the left side of the beam is shown. Values for the two sides of the beam are averaged to yield the figures in the table. The upper beam occurs above the aiming point and the lower beam below it. This data applies no matter where the luminaire is aimed in the lighted area.

Each block represents a section of the beam. Each subtends an equal angle horizontally and vertically away from the floodlight's aiming point. The horizontal and vertical degree increments are the same for symmetrical floodlights, but two different values may be used for asymmetric floodlights. The degree increments are selected to cover most of the lamp lumens emitted by the floodlight. The data usually extends well beyond the one-tenth of maximum candela point, which is the edge of the beam and is shown by the dashed line. This makes the design more accurate than it would be if only lumens in the beam were included.

There are nine blocks horizontally and vertically in the two sections of the left side of the beam: the left upper and left lower. This does not include the blocks in the 0-degree reference lines. In the example, each block contains two values: the upper number is the luminous intensity in candelas in the center of the block or the middle of that portion of the beam; the lower number indicates the total number of lumens which fall within the section of the beam bounded by the block. Both are based on 1000 lamp lumens. These values, particularly the candela value, can be used in floodlight design. This is explained in the next chapter.

At the left side of the photometric chart is a summary of the total lamp lumens included in the table. As shown in Fig. 9-1, 63.2 percent of the lamp lumens are represented: 32 percent above and 31.2 percent below the aiming point. However, the beam (which is the area inside the dashed line) only includes 52 percent of the lamp lumens. This figure can be read in the lower left portion of the data and is called *beam efficiency*. It is the percentage of the lamp lumens that falls within the beam. The beam efficiency goes down as the beam spread becomes narrower, even though the total lumens radiated from the luminaire may remain essentially the same. Low beam efficiency does not necessarily mean that the floodlight is inefficient. Beam efficiency or total efficiency must be multiplied by the utilization factor to determine system efficiency. This is covered in the next chapter.

Candela Traces

To the right of the candela and lumen distribution charts in Figs. 9-1 and 9-2 is a graph called the *candela traces*. This shows the pattern of the candelas that are being projected from the floodlight at various angles, both horizontally and vertically, from the maximum candela point to the one-tenth of maximum candela points. These graphs are based on candelas per 1000 lamp lumens. For symmetrical floodlights, sometimes only one trace is given, since the distribution is similar in all directions from the axis of the reflector and values can be averaged. For asymmetric floodlights, there are always two curves: a horizontal trace and a vertical trace through the centerline or aiming axis of the floodlight. For asymmetric floodlights the upper vertical beam is shown on the right side of the graph and the lower vertical beam on the left.

Rectilinear coordinates are used for the candela values rather than polar coordinates, which would be more representative of the actual beam shape. This makes it easier to read the change in the candela distribution at the edge of the beam.

Floodlight Classification

Floodlights are classified on the basis of their horizontal and vertical beam spreads, as shown in the candela traces. The beam spread is the total number of degrees between the one-tenth of maximum candela point on one side of the beam and the one-tenth of maximum candela point on the opposite side. In symmetrical floodlights, the maximum candelas almost always occur on the axis of the reflector, which is also the aiming axis. In asymmetric floodlights, the maximum candelas can occur above the aiming axis or to either side of it. In any case, the one-tenth of maximum candela point is measured from the maximum candela point, regardless of where it falls.

There is one beam spread for the horizontal beam, and another for the vertical. These beam spreads are divided into numbered types by the National Electrical Manufacturers Association (NEMA), as shown in Table 9-1. The lower the NEMA number, the narrower the beam spread.

Table 9-1 Outdoor floodlight luminaire designation

Beam Spread, degrees	NEMA Type Designation
10 to 18	1
18 to 29	2
29 to 46	3
46 to 70	4
70 to 100	5
100 to 130	6
130 and up	7

SOURCE: Reproduced with permission from *Outdoor Floodlighting Equipment,* NEMA pub. no. FA 1-1973, National Electrical Manufacturers Association, New York, 1973.

Each number covers a range of beam spreads. Floodlight photometric data should always include the NEMA number of the floodlight.

If the floodlight beam is symmetrical, the horizontal and vertical beams are so similar that their beam-spread values are averaged and the same NEMA number given for both the horizontal and vertical beams. For instance, in Fig. 9-1 the lower left corner shows that the NEMA type for the floodlight is 5 by 5, and that the beam spread is 92 degrees horizontal and 89 degrees vertical. (The NEMA numbers 6 by 6 are the values for this floodlight under the older method of evaluating photometric data.) Note on the candela trace that this means the beam extends 46 degrees on either side of the maximum candela value of the floodlight.

If only one NEMA number is assigned, the floodlight is symmetrical and the candela trace represents the average in all directions from the axis. Asymmetric floodlights usually have a beam that is wider horizontally than vertically and are assigned two NEMA numbers, with the horizontal beam spread always given first. In Fig. 9-2, the candela trace shows that the one-tenth of maximum candela point for the horizontal beam falls 55 degrees either side of the aiming direction, for a beam spread of 110 degrees. Table 9-1 shows that NEMA 6 includes floodlight beam spreads from 100 up to 130 degrees. The vertical beam extends from 41 to 42 degrees, for a total beam spread of 83 degrees, which falls within the NEMA 5 classification.

Floodlights should never be selected on the basis of their beam spreads in degrees or NEMA numbers alone. Candela traces should be used to check the actual shape of the beam. Uniformity will suffer if floodlights have irregularly shaped beams. Hockey-stick changes in candela distribution such as those shown in Fig. 9-3 have a measured beam spread wider than that which the floodlight actually produces. A more accurate value for the effective beam in this case would be given by the dashed line. The dashed line would indicate a spread of 50 degrees and a NEMA 4 rating rather than a spread of 74 degrees and a NEMA 5 rating.

Floodlight Description

The Floodlight Description in the examples (Figs. 9-1 and 9-2) is given in the lower right corner. This includes information about the lamp type used in the floodlight, possible part numbers, and the photometric test or curve number. Stated here also would be whether there are any modifications which would change photometric performance, such as the addition of visors, louvers, or shields. The NEMA description may also be included.

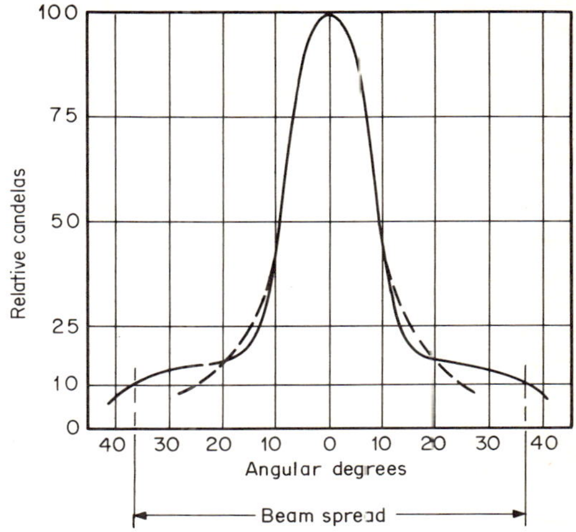

Fig. 9-3 When a floodlight beam spread curves sharply outward at the perimeters, the NEMA number designating that beam spread may be misleading. The NEMA number for this floodlight would be 5, whereas the dashed line and a NEMA 4 rating would give a more accurate picture of the beam spread.

Isoillumination Diagrams

Published photometric data is based on a floodlight aimed at a right angle to the surface being lighted. Since floodlights are normally aimed at an angle to the surface, other methods of presenting the photometric performance are helpful.

The most widely used are isoillumination diagrams. These are either isofootcandle or isolux diagrams, depending on whether the data is presented in footcandles or lux. These diagrams are available from some manufacturers.

Individual isoillumination diagrams are prepared for each floodlight type at several different aiming points. Figure 9-4 gives a set of isoillumination diagrams for the floodlight shown in Fig. 9-2. The black dot indicates the point where the floodlight is located, and the arrow shows the aiming point.

Isoillumination diagrams can be drawn for a specific mounting height, with any convenient scale for the coordinates. Another, more flexible method uses a grid scale which is equal to the mounting height. In the second case, the diagrams can be used for any mounting height,

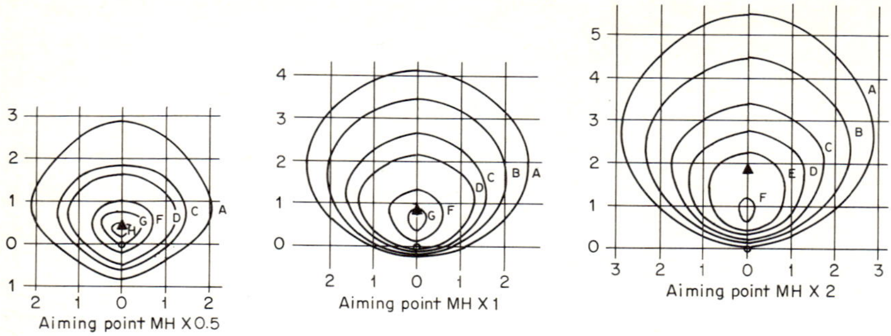

1000-watt high-pressure sodium

	Mounting Height, 15.2 meters (50 feet) Lux	Fc
A	1.1	0.1
B	2.2	0.2
C	5.4	0.5
D	10.8	1.0
E	21.5	2.0
F	54.0	5.0
G	108.0	10.0
H	161.0	15.0

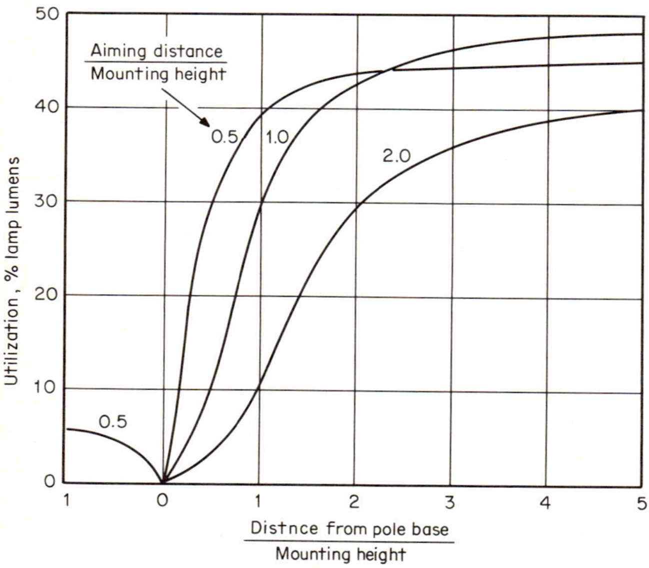

Fig. 9-4 Isoillumination diagrams and utilization curves for the HPS floodlight in Fig. 9-2. The lines in the diagrams connect points of equal illumination. Actual illumination levels vary with mounting height and are read from accompanying charts, such as the one given.

expressed in either feet or meters. For a 40-foot mounting height the distance between grid lines would represent 40 feet, for a 10-meter mounting height they would represent 10 meters, and so on.

The shape of the isoillumination lines remains the same as the mounting height changes, but the illumination values they represent vary as the square of the mounting height. As the mounting height increases, the size of the area that is lighted increases as well. Since the lumens emitted by the floodlight remain constant, the lumens per unit area decrease and so does the level of illumination. Conversely, if the mounting height decreases, the level of illumination increases.

Each line on the isoillumination diagram connects the points which have the same initial illumination value, whether measured in footcandles or lux. The separation between isoillumination lines should be such that there is approximately a 2:1 ratio in illumination level between lines. Using a 2:1 ratio allows the values between lines to be interpolated without significant error.

Tables giving the lux or footcandle values associated with each line at different mounting heights are given with each set of isoillumination diagrams. Figure 9-4 includes such a table for a 1000-watt high-pressure sodium (HPS) floodlight at a mounting height of 15.2 meters (50 feet). For other mounting heights, use the formula, existing $E \times MH^2 =$ new $E \times mh^2$, where E is the level of illumination and MH (mh) is the mounting height. The illumination values for each letter in Fig. 9-4 are the same on each set of curves for any one lamp. This makes it possible to compare diagrams directly and interpolate between curves.

The centermost line in any diagram should give the maximum initial illumination value that the floodlight will produce at the specified aiming angle and mounting height. For good uniformity this value should be close to the desired average initial level of illumination. Higher values are possible but may result in visible hot spots near the floodlight if the maximum value is over 3 times the average.

The same isoillumination diagram can be used for several lamp sizes if tests indicate that the shape of the beam remains the same or changes very little between lamp sizes. In the example, Fig. 9-4, the same three diagrams apply to the 250-watt, 400-watt, and 1000-watt HPS lamps. However, the tables of illumination levels associated with each lamp are different.

Just as the candela trace makes it easy to compare the beam patterns of different floodlights, the isoillumination curves can be used to compare the coverages that different floodlights provide. The designer can also use them to determine the maximum horizontal separation between aiming points, by superimposing the edges of isoillumination diagrams over each other. The horizontal illumination at the intersection of two isoillumination lines is the sum of the values given for each line and should equal the desired minimum or average light level.

Generally, at least two floodlights are mounted at each location if they are perimeter-mounted, in order to cover 180 degrees. If the floodlights are mounted within the area, at least four are needed to cover 360 degrees. If more than one floodlight is to be mounted at a location, the isoillumination diagrams can be drawn to the same scale as the working drawings, and copies can be made. Isoillumination diagrams for more than one floodlight type can be used. The designer can then position these on the working drawings, matching isoillumination lines until the desired levels of illumination are obtained and the area is completely covered.

Utilization Data

Some manufacturers publish utilization curves with each set of isoillumination diagrams. The information provided is the same as the summary given in the candela and lumen distribution chart in the photometric data; but it has been recalculated for the specified aiming angle. These curves show the percentage of initial lamp lumens that falls in each horizontal area measured from the base of the floodlight location. They are based on the mounting height of the floodlight and the aiming distance, which is expressed as a multiple of the mounting height (MH). The number beside each curve gives the aiming point so that the utilization curve can be identified with the associated isoillumination diagram. This method of determining floodlight utilization is similar to the method used for roadway lighting design.

Utilization data applies to horizontal sections which are perpendicular to the aiming line of the floodlight. There will be some loss of accuracy if perimeter-mounted floodlights are not aimed directly into the area being lighted. Utilization data still applies if the approximate edge of the beam falls within the area being lighted. The approximate edge of the beam is sometimes noted on the isoillumination diagrams by a dashed line. If there is any question whether or not the edge of a floodlight beam will fall within the lighted area, the appropriate isoillumination diagram can be redrawn to the scale of the working drawings and used as a template to show exactly where any given beam will fall.

Utilization curves give information on both uniformity and utilization, thus providing another tool to use in selecting the proper aiming distance. Figure 9-4 includes the utilization curves for the three isoillumination diagrams. The approximate uniformity for each aiming point can be determined from the graph. For example, assume that an area 3 MH wide must be lighted. In Fig. 9-4 the curve for the 2-MH aiming point puts 10 percent of the lamp lumens from 0 to 1 MH, an additional 20 percent from 1 to 2 MH, and 5 percent from 2 to 3 MH.

Again reading from Fig. 9-4, the 1-MH aiming point puts 30 percent of the lamp lumens from 0 to 1 MH 13 percent from 1 to 2 MH, and 3 percent from 2 to 3 MH. Therefore this floodlight, provides better uniformity for this area when aimed at 2 MH. Uniformity will be different if more than one floodlight is mounted at the same location.

The best uniformity would be indicated by a utilization curve that is a straight line from the base of the pole to the point on the graph indicating the opposite side of the lighted area, for one-side lighting, or the middle, for lighting from both sides. This would provide poor utilization, however, since much light would be lost beyond the area that had to be lighted.

Floodlights that light an area within 2 MH of the pole can have an average lamp-lumen utilization of 50 percent; for areas up to 3 MH this drops to 40 percent; and for areas wider than 3 MH it is only 30 percent. As might be expected, utilization goes down as the floodlight is aimed farther from the base of the pole. The highest possible utilization occurs when the aiming line (or maximum candela line, for asymmetric floodlights) bisects the angle made at the floodlight by the near and far sides of the area being lighted. If an area is lighted from one side, this point will always occur at a point closer than halfway across the area being lighted. Aiming at this point, however, will invariably produce poor uniformity.

Selecting an aiming point, then, must be a compromise between uniformity and utilization. The ideal utilization curve would rise steeply to the line on the graph indicating the furthest distance the floodlight had to cover and would then flatten out. An aiming point of a little over 2.5 MH at a vertical angle of 70 degrees seems to be the practical limit for floodlights lighting a horizontal surface. A very narrow beam floodlight can light a flat, unobstructed area out to 5 MH, but with a great loss of both utilization and uniformity.

Vertical Aiming Angles

Floodlights are almost always aimed at an angle to the ground. It is relatively simple to light an area out to 1 MH, because the floodlight has 45 degrees in which to distribute light. An additional 18.4 degrees are available from 1 to 2 MH. From 2 to 3 MH only 8.2 degrees are available, and from 3 to 4 MH only 3 degrees. This is illustrated in Fig. 9-5.

In other words, as the distance from the pole increases, the area the luminaire "sees" gets larger and larger in proportion to the vertical angle within which the floodlight must project the needed light.

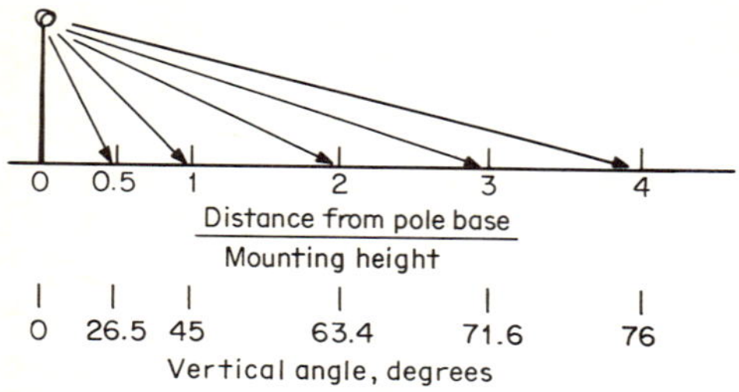

Fig. 9-5 When floodlights are aimed close to the pole, a greater angle is available in which to spread light. It becomes more and more difficult to light an area as the aiming point is moved farther from the pole base.

Horizontal levels of illumination go down as aiming angles are increased. The ray D in Fig. 9-6 can be divided into the vectors V and H, where V is the vertical component and H the horizontal. The illumination level E produced at any given point can be obtained by the formula

$$E = \frac{\text{cd} \cos \theta}{D^2}$$

where E = level of illumination
cd = candlepower expressed in candelas
$\cos \theta$ = cosine of angle θ
D = distance from floodlight to aiming point

Since it is difficult to measure D directly, other approaches are generally used. If the length of D is proportional to the candelas of the luminaire, then V is proportional to the horizontal illumination that the ray D will produce at point P, and H is proportional to the vertical illumination that the ray D will produce at point P, perpendicular to a line drawn from the base of the pole. Values, then, can be expressed in ratios:

$$\cos \theta = \frac{\text{V}}{\text{D}} \qquad \text{or} \qquad \text{D} = \frac{\text{V}}{\cos \theta}$$

Substituting in the original formula,

$$E = \frac{\text{cd} \cos \theta}{(\text{V}/\cos \theta)^2} = \frac{\text{cd} \cos^3 \theta}{\text{V}^2}$$

As seen in Fig. 9-6, V = MH, and so

$$E = \frac{\text{cd} \cos^3 \theta}{\text{MH}^2}$$

Thus the illumination at any point away from the pole can be expressed as a function of the cosine cubed of the incident angle θ. A more convenient method is to express the number of candelas required in order to produce a given illumination level at a distance from the pole base as a function of the illumination produced at the base of the pole and the distance from the pole base. The table in Fig. 9-6 is based on the latter method. The distance from the pole base to point P is H/V, which can be expressed in mounting heights (V).

At the base of the pole the cosine of the incident angle θ at P is 1, so

$$E = \frac{\text{cd} \times 1}{\text{MH}^2}$$

The pole height used to tabulate the chart was 31.6 feet (9.6 meters), which is the square root of 1000. The desired illumination leve. is 1 footcandle (10.76 lux), and so

$$1 = \frac{\text{cd}}{\text{MH}^2} = 1\times 31.6^2 = 1000 \text{ candelas}$$

to produce 1 footcandle at the base of a 31.6-foot pole.

Any pole height could have been used, but if the candela value at the base of the pole is a multiple of 10 other values can be calculated directly with the ratios in the last column in Fig. 9-6. These ratios are valid for meters and lux as well as feet and footcandles.

To produce the same illumination level at the aiming angle of 63.4 degrees, or 2 MH (2 × 9.6 = 19.2 meters) from the base of the pole, would require about 11.2 times as many candelas, or 11.2 × 1000 = 11,-200 candelas.

For a 30.48-meter (100-foot) mounting height, the candelas needed in

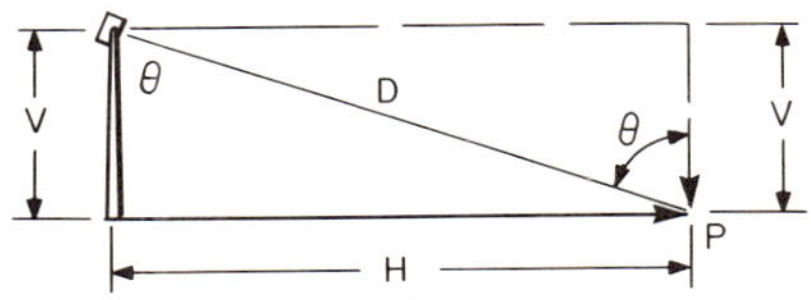

H/V	*θ*	Candlepower Expressed In: Cosine³ *θ*	Candlepower Expressed In: Candelas	Ratio
0	0	1.0	1,000	1.0
0.5	26.6	0.715	1,398	1.4
1.0	45.0	0.3536	2,828	2.8
1.5	56.3	0.1707	5,859	5.86
2.0	63.4	0.0894	11,180	11.2
2.5	68.2	0.0512	19,521	19.5
3.0	71.6	0.0316	31,623	31.6
3.5	74.05	0.0207	48,230	48.2
4.0	75.96	0.0143	70,093	70.1

Fig. 9-6 A light ray such as D can be divided into horizontal and vertical components (H and V). The table indicates the number of candelas needed to produce 10.76 lux (1 footcandle) at various ratios of H to V from a mounting height of 9.6 meters (31.6 feet).

order to produce 10.76 lux (1 footcandle) at the base of the pole would be $10.76 \times (30.48)^2 = 9997$, or approximately 10,000 candelas ($1 \times 100^2 = 10{,}000$). At 2 MH, the ratio would remain the same as in the previous example—11.2 times as many candelas would be needed—and so $11.2 \times 10{,}000 = 112{,}000$ candelas would be needed to produce 10.76 lux (1 footcandle) at a distance 2 MH from the base of the pole.

In the case of an area 2 MH wide it is possible to use only one type of luminaire to light uniformly back to the base of the pole. This is true because the 10:1 candlepower ratio needed to cover the width of the area falls close to the maximum-candela-to-one-tenth-candela pattern of a floodlight beam.

With aiming points beyond 2 MH, however, candlepower requirements rise so rapidly that it is difficult to produce a uniform lighting system with only one type of floodlight. From the chart, it can be seen that it takes 70.1 times as much candlepower to produce 10.76 lux at 4 MH as it does at the base of the pole, and 31.6 times as much at 3 MH. This means that more than one type of luminaire would be needed in either case: one with a very narrow beam to concentrate candlepower and project it a great distance, and another with a wide beam to fill in the area close to the pole.

Luminaire Selection

The designer must understand these ratios in order to pick a luminaire on the basis of its maximum candlepower, which is luminous intensity expressed in candelas. The maximum number of candelas that any floodlight produces can be obtained from photometric data.

The maximum candela value is a function of the lamp brightness and the flashed area of the reflector. The flashed area of the reflector is controlled by the reflector design. A narrow-beam reflector will be fully flashed when viewed along its centerline or aiming axis. Floodlights are usually too bright to be viewed directly, but the flashed area can easily be seen if viewed through a filter. In a narrow-beam floodlight the maximum candela value will not increase with higher-wattage lamps unless the increased wattage results in a brighter lamp. The higher wattage will, however, increase the beam spread if the increased wattage is achieved with a larger diameter or longer source size.

Floodlights come in families of lamp sizes, beam spreads, and maximum candela values. Lamp size determines the total lumens emitted by the floodlight. Reflector design determines the maximum candela value and beam spread.

The selection of a particular floodlight by its maximum candlepower cannot be too precise and does not need to be. The point where this candlepower strikes the surface may not be the brightest spot, but its brightness is easily measured and gives a good indication of general uniformity. The main objective is to ensure that the maximum candlepower does not produce a spot of light within the lighted area that is significantly brighter than the average level. If this happens the lighting system will be visibly nonuniform. It is best to keep the maximum light level produced by each floodlight as close to the average light level as possible, or at least not more than 3 times the average value.

It is important also not to make the illumination level at this spot lower than the average level. If a floodlight has significantly less candlepower than is required in order to provide the desired light level its size is probably too small or its beam too wide. In both cases the system will be inefficient: in the first case because more floodlights will be needed and in the second because the wider beam will probably throw more light than necessary beyond the lighted area.

Angle of Maximum Candlepower

Unless a floodlight is aimed straight down at the surface being lighted, the maximum level of illumination will always occur behind the point where the floodlight is aimed. The wider the beam spread of the luminaire, or the higher the aiming angle, the farther this maximum level of illumination will occur behind the aiming point. There is one aiming angle, however, which will produce the highest level of illumination at a given distance from the pole. This occurs at an aiming angle of 54.7 degrees. If the maximum candela point of the floodlight is aimed at this angle, the highest level of illumination the floodlight is capable of producing away from the pole will be obtained. The level of illumination will be higher behind the aiming point, but this will give the maximum level of illumination at the greatest distance from the pole.

Raising the pole height, or raising or lowering the vertical aiming angle, will reduce the illumination value at this point. Figure 9-7 shows a floodlight aimed at an angle of 54.7 degrees. The triangle formed is almost in a 3:4:5 ratio. The mounting height (MH) is 3, the horizontal distance on the surface (*H*) is 4, and the distance (*D*) from the floodlight to the aiming point P is 5. This means that mounting height is three-fourths of the distance from the pole to the aiming point (MH/H = 3/4), or the aiming point is a distance of about 1⅓ MH.

If there is a particular part of an area away from the pole that needs to receive the highest possible level of illumination, the floodlight mounting height should be selected so that the floodlight can be close to an angle of 54.7 degrees.

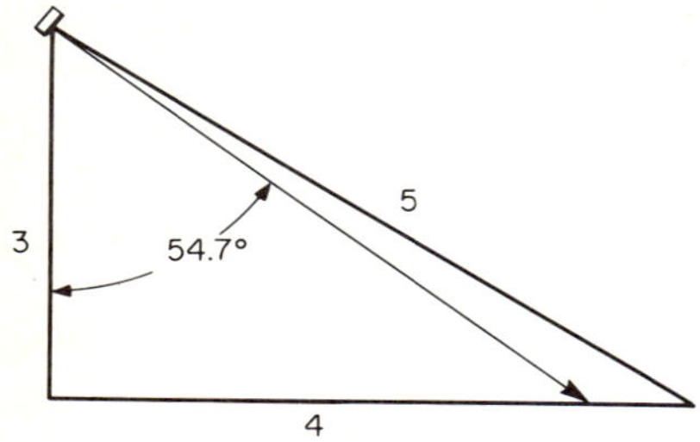

Fig. 9-7 Aiming the maximum candlepower of the floodlight at 54.7 degrees will produce the highest possible level of illumination at a distance from the pole.

Horizontal Aiming Angles

The horizontal aiming angle of a floodlight is the angle at which the floodlight is aimed in relation to a line drawn from the base of the floodlight location at a 90 degree angle to the perimeter of the area. The best method of selecting the horizontal aiming angle is to use the isoillumination diagrams. Floodlights should be aimed at the angle which gives the best coverage of the area.

The approximate horizontal separation between aiming lines can also be determined from the photometric candela trace. For the most uniform system the overlap between adjacent beams should occur no higher than the one-half of maximum candela point. For the symmetrical floodlight in Fig. 9-1, this point is 552 candelas and occurs at approximately 11 degrees to each side of the aiming line. This means that the maximum horizontal angular separation between floodlight aiming lines for this floodlight should be approximately 22 degrees.

For a 2:1 change in illumination level between aiming lines the one-quarter maximum candela point (276 candelas) can be used. This occurs at approximately 28 degrees, and so the aiming-line separation can be as much as 56 degrees.

Computer services offered by some manufacturers can give point-by-point illumination levels for a floodlight aimed at various angles. This can help to determine aiming angles.

Often, if a job is small or uncomplicated, the simplest thing to do is to estimate the horizontal aiming angle and to make adjustments in the field.

Vertical Illumination

In floodlight design, the vertical level of illumination is just as important as the horizontal. Many things that need to be seen in a work or storage area are above the surface, such as the sides of boxes, freight cars, etc. If the horizontal level of illumination (E_H) at a point is known, the vertical level in line with the floodlights at the same point can be calculated directly. The ratio of vertical to horizontal illumination is the same as the ratio of the horizontal distance from the base of the pole (H) to the mounting height (MH):

$$\frac{E_V}{E_H} = \frac{H}{\text{MH}}$$

For example, at a distance of 2 MH from the base of the pole the vertical level of illumination is twice the horizontal ($H/\text{MH} = 2$), and at 4 or 5 MH the vertical illumination level is 4 or 5 times the horizontal. This simple relationship works only if light is coming from a single location. If several poles are involved, the contribution from each has to be calculated individually and the results added.

Shadows

Shadows are another factor to consider in design. For a single luminaire, the length of a shadow is directly proportional to the height of the luminaire and the distance to the pole base (Fig. 9-8). At a horizontal distance equal to twice the mounting height, the length of the shadow will be twice the height of the object casting it. This can be shown by the formula

$$\frac{\text{MH}}{H} = \frac{D_1}{D_2}$$

where MH = the luminaire mounting height
H = the height of the object casting the shadow
D_1 = the distance from the pole base to the end of the shadow
D_2 = the length of the shadow

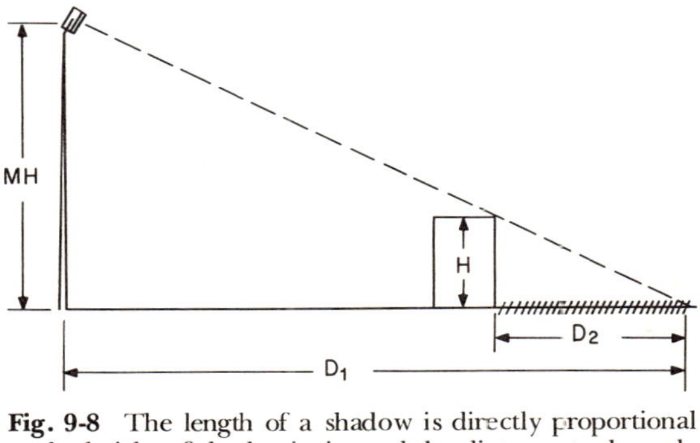

Fig. 9-8 The length of a shadow is directly proportional to the height of the luminaire and the distance to the pole base. Therefore $MH/H = D_1/D_2$.

The higher the poles are, the shorter the shadows will be. This can be an important consideration in an outdoor work area crowded with equipment and materials. Shadows can be reduced by providing light from more than one direction.

Bibliography

"IES Approved Method for Photometric Testing of Floodlights Using Incandescent Filament or Discharge Lamps," *Illuminating Engineering,* February 1971, pp. 107–121.

Outdoor Floodlighting Equipment, NEMA standards pub. no. FA 1-1973, National Electrical Manufacturers Association, New York, 1973.

10

Design for Floodlighting Systems

The designer needs a thorough understanding of photometric data before designing a floodlighting system. Acceptable results can be obtained without this knowledge only if there is little need for uniformity or minimum light levels have not been specified.

The preceding chapter, Chap. 9, gives a detailed analysis of how to evaluate the performance of individual floodlights. This chapter will discuss an actual design problem and all the factors that go in to a final decision on system design.

Eye Adaptation

The first step in the design process is to look at the actual situation in which a worker will be expected to perform certain tasks. Our eyes enable us to see in almost any conditions, from starlight to full sunlight. What the eye sees is luminance or brightness—not the illumination level, which is easy to measure in footcandles or lux. The total field luminance or average brightness of the surround affects visibility, since it controls the level of eye adaptation. This includes the brightness of both vertical and horizontal surfaces, although only horizontal levels of illumination are usually included in specifications. Vertical levels of illumination are especially important in floodlighting, because often the area to be lighted is a storage yard or work area where objects above the ground, as well as horizontal surfaces, must be visible.

The size of the objects to be seen out of doors at night is important, too, as well as the amount of time available for the person to adapt to the outdoor lighting. Eyes can adapt to a range of 10^{12} change in brightness, from shadows in starlight to the brightness of snow in full sunlight, but a certain amount of time is necessary.

Tests have shown that the human eye can take as long as 3 to 5 seconds to adjust to a change in illumination from 323 to 430 lux (30 to 40 footcandles) to moonlight (0.1 lux, or 0.01 footcandles). This would be for detail the size of the largest letters in a standard Snellen eye-test chart. For older people adaptation times can be as much as twice as long.

An adaptation period of several seconds is not well suited to industrial applications. However, the eye can adapt almost instantly to a 100:1 change in luminance. If a worker is required to go from a factory to a storage yard and the factory light level is 1076 lux (100 footcandles), then the light level in the outside area can be 1/100 times lower, or 10.76 lux (1 footcandle). The worker will easily adapt to the change in light level and be able to find fairly large items.

If detailed work must be done out of doors (for example, reading fine print on an invoice), it would probably be more economical for the worker to carry a flashlight than for the designer to light the whole yard. Portable equipment could also be available. A light on the front of a lift truck, for instance, might provide enough illumination for the needed task.

Levels of Illumination

When choosing the appropriate level of illumination for a given situation, the designer must consider both the work to be performed outdoors and the adaptation requirements. Some government agencies are now recommending that certain light levels be met, and these must also be considered. The Occupational Safety and Health Administration (OSHA) requires a minimum of 5 footcandles at any point when earth-moving equipment or front loaders are used at night, and 3 footcandles for "general construction areas, loading platforms, refueling, field maintenance areas." These specific levels are chosen for safety.

Other levels of illumination recommended by the Illuminating Engineering Society of North America (IES) are included in Chap. 1. These levels may be different if the area must satisfy the needs of closed-circuit television.

For a sample problem, assume that an outdoor work area 54.9 by 54.9 meters (180 by 180 feet) must be lighted. It is an unloading area in a shipyard, which requires a light level of 54 lux (5 footcandles).

Placement of Lighting Equipment

After the light level, the next consideration is the location of luminaires and poles. This placement should be coordinated with work areas, the flow of traffic, and the presence of obstructions. Lighting equipment can be mounted on poles, buildings, high structures, equipment, or any combination of these. Sometimes floodlights are mounted directly on a crane used to pick up stored items, or on a conveyor belt.

In extremely congested areas a catenary system can be used. Luminaires are suspended on a cable strung across the lighted area. Catenary systems are designed in the same manner as other floodlighting systems. In some cases modified indoor lighting equipment is used. This equipment is aimed straight down, and so indoor design procedures are followed.

In deciding on luminaire placement, the designer must also consider appearance and glare. Persons coming into the area should not be faced with glaring luminaires. If there are community regulations affecting pole height and/or glare, these should be considered. Common sense plays a part also; if there is an attractive building in the area, floodlights should not be placed on top of it, because the brightness of the lights would obscure the details of the building.

For the example, assume that there are no restrictions on pole height or placement, and no community regulations regarding glare levels.

Perimeter Lighting

Lighting equipment can be placed within an area or around the perimeter. If poles are located within the area, parking-lot or high-mast luminaires are a more efficient choice than floodlights. The designer should consider these first, to determine whether they can be used. Information on their use is given in Chaps. 11 and 12.

Perimeter lighting is the most common method of utilizing floodlights. This placement does not restrict activity within the area and also can provide light from more than one direction, which contributes to good uniformity and lessens the problem of shadows.

There are three typical ways of lighting a square or rectangular area. In the first, floodlights are mounted in each of the four corners so that light is provided throughout the area from four different directions. This minimizes shadows and is also the most efficient method to use, since most of the light falls within the area. Often, only one floodlight has to be mounted on each pole; when other mounting positions are used two are usually needed,

In the second method, floodlights are mounted opposite each other and midway along each side of the area. This method is only slightly less efficient than the first. Often it is a more economical approach, since fewer poles are needed, For the small area in the example, only two poles would have to be used, with two floodlights mounted on each pole.

In the third method, floodlights are mounted in one location only, usually midway along one side of the area. This is the least satisfactory of the perimeter approaches, but it can produce satisfactory results if more than one floodlight is used so that light from adjacent floodlights will overlap to cover the area completely. Even so, it will be difficult to light the corners and far side adequately, and so this approach cannot be used in a situation where there is activity throughout the area. There can also be a problem with shadows, since light comes from only one direction.

Pole Height and Spacing

The size and shape of the area usually dictate the height and spacing of poles; a rough layout of the area should therefore be made. Figure 10-1 is the preliminary layout for the sample problem. For more complicated jobs, the area can be divided into more than one section for design purposes.

If the area is to be lighted from only one side, pole height, or mounting height, should be approximately one-half the width of the area to be lighted. In this way, the floodlights can be aimed almost to the other side of the area, or at an aiming distance of just under 2 times the mounting height (2 MH). This provides the best uniformity while still producing reasonable utilization, as explained in the previous chapter. For larger areas, pole spacing should be 4 MH, since each floodlight can efficiently light an area up to 2 MH wide.

If the sample area, which is 54.9 by 54.9 meters (180 by 180 feet), were to be lighted from one side, a pole 27.4 meters (90 feet) high could be used. If the same area were to be lighted from more than one location, 13.7-meter (45-foot) poles would be a better choice. This would enable the floodlights to be aimed at just under 2 MH or a little less than 27.4 meters, into the center of the area. If the floodlights were mounted in the four corners, they could still be aimed at 27.4 meters, because the center of the square would be lighted with the contributions from all four corners.

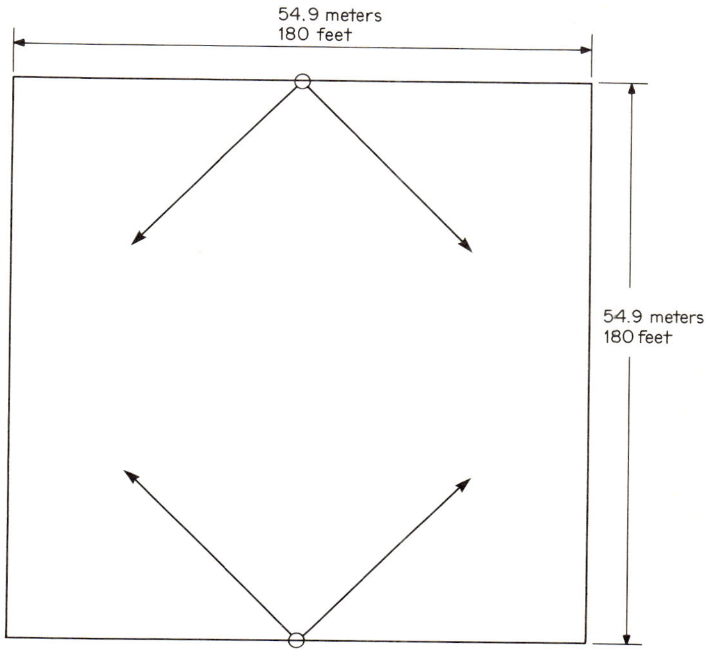

Fig. 10-1 Layout of the sample outdoor area.

For the purposes of this example, assume that the 54.9- by 54.9-meter area will be lighted with floodlights mounted on 13.7-meter poles opposite each other and midway along the edge of the area to be lighted. A minimum of two floodlights will be needed on each pole to cover the area, and so a total of four floodlights will be needed. This is an estimate only, and must be checked through the design procedures that follow.

Types of Lamps

Before deciding which floodlight to use, it is a good idea to investigate the performance of various light sources. Floodlight efficiency does not change significantly between sources, and so the efficacy, or lumens per watt, delivered by the light source will determine system efficacy. Detailed information on light sources is given in Chap. 2.

High-pressure sodium (HPS) sources provide good efficacy and will be used to light the sample area. A floodlight with an asymmetric beam pattern will be used because the wider horizontal beam will give better coverage for this application than a circular beam pattern would.

Lumen Method

The lumen method is the most common way to estimate the number of lamp lumens needed to light a particular area. From this information, the designer can estimate the number of floodlights needed. All that is really known, however, is the approximate average illumination level that will be provided. This is necessary preliminary information, but it gives no indication of how well the light will be distributed in the area. To check for uniformity, the designer should use one of the methods for checking point-by-point values that are given later in this chapter.

The following formula is used to solve for lamp lumens:

$$\text{LL} = \frac{E \times \text{area}}{\text{CU} \times \text{LLF}}$$

where LL = lamp lumens
E = initial illumination level in lux or footcandles
CU = coefficient of utilization
LLF = light-loss factor

The designer can use Fig. 10-2 in place of the formula by following this procedure:

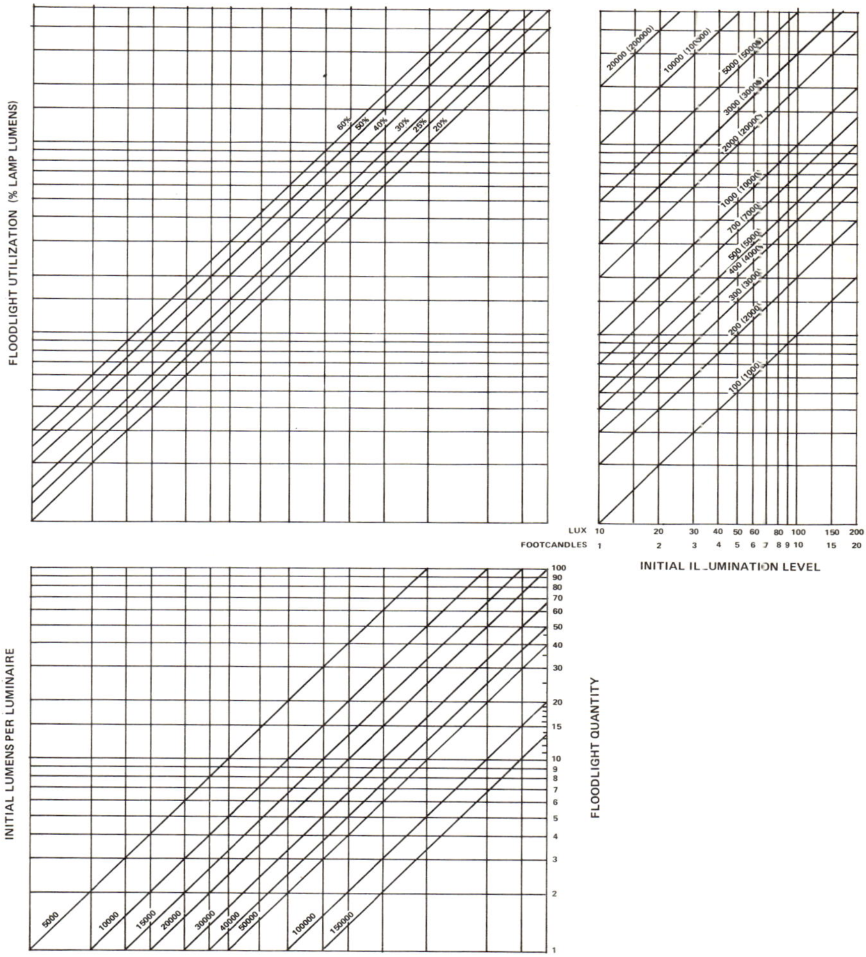

Fig. 10-2 Chart used to select the correct floodlight and the number that will be needed. It must be followed in one measurement system all the way through.

STEP 1

Select the type of light source that will be used for this application and determine the necessary maintained light level. In the example, floodlights containing HPS lamps will be used. A 1000-watt lamp will be tried first, since it is always best to try to use the highest wattage possible in order to use as few luminaires as possible.

Recommended maintained illumination levels can be read in Chap. 1 or found in the *American National Standard Practice for Industrial Lighting.*

The shipyard used as an example requires a maintained level of 54 lux (5 footcandles).

STEP 2

Use the light-loss factor to find the initial recommended light level. Light-loss factor consists mainly of lamp lumen depreciation (LLD) and luminaire dirt depreciation (LDD).

From manufacturer's published lamp and luminaire information, find the LLD and LDD factors. The LLD factor gives an indication of how much of the rated output of the lamp will still be produced at the mean point in the lamp's life. (The end-of-relamping figure should be used for television-surveillance lighting.) For the example, 0.9 is the LLD for the 1000-watt HPS lamp.

The LDD takes into account the effect of dirt on the luminaire's light output. The luminaire LDD which will be used for the example is 0.95, for a sealed and filtered floodlight. (See Chap. 3 for luminaire characteristics.)

For some sources, such as mercury and metal halide, a tilt factor must also be considered; initial lamp lumens must be determined on the basis of the lamp's operating position. The figure for tilt factor can be obtained from manufacturer's lamp data.

If the desired maintained light level is known, the initial level can be determined from the following formula:

$$\frac{\text{Lux or fc (maintained)}}{\text{LLD} \times \text{LDD} \times \text{tilt factor}} = \text{lux or fc (initial)}$$

High-pressure sodium luminaires do not require a tilt factor. Substituting in the formula for the sample problem,

$$\frac{54}{0.9 \times 0.95} = \frac{54}{0.855} = 63.2 \text{ lux}$$

Levels of illumination in either lux or footcandles can be used for Fig. 10-2, but the graph must be followed in one measurement system all the way through. For the example, lux and meters will be used.

Find the initial light level of 63 lux on the bottom ordinate of the right section of the graph.

STEP 3

Refer to the preliminary layout of the location of the floodlights (Fig. 10-1) and calculate the size of the area to be lighted. The 54.9 by 54.9-meter area in the example is 3014 square meters.

Draw a line from the initial light level up to the 3000 square meter line.

STEP 4

Estimate the coefficient of utilization (CU) for the lamp-floodlight combination that has been chosen. The CU is the percentage of the lamp lumens that fall into the lighted area and contribute to the light level. (The coefficient of beam utilization is used when beam lumens are used in place of lamp lumens—a method that will not be used here.)

Table 10-1 gives estimated efficiencies for various area sizes and pole locations. These are first approximations only and must be adjusted when the performance of the selected floodlight is known. Table 10-1 assumes that poles are on the perimeter of the lighted area. If they are to be set back, the lighted area should be extended to the base of the pole and a new figure for the total area used in following the graph.

In the example, with poles on two sides and a width of 4 MH (4 × 13.7 = 54.8 meters) the utilization would be 40 percent. Draw a line from the 3000-square meter point to the 40 percent utilization line in the upper left portion of the graph. From there, draw a line straight down into the last section of the graph, where the lamp lumens are listed.

STEP 5

In the example, the number of floodlights has been estimated. The tentative floodlight quantity is four floodlights, and so a line can be drawn from the spot that indicates four floodlights, on the right of the graph, to meet the line drawn down from the utilization point. The lines meet between the 100,000-lumen line and the 150,000-lumen line; a lamp rated somewhere between these values could therefore be used. If a lamp rated at 100,000 lumens were used, five luminaires would be needed. This can be read from the graph by drawing a line from the 100,000-lumen line to the ordinate indicating luminaire quantity. Three luminaires would be adequate if a lamp rated at 150,000 lumens were used.

STEP 6

Refer to Table 2-1 in Chap. 2 to see which types of lamps emit the required number of lumens. The 1500-watt metal halide lamp is rated at

Table 10-1 Estimated floodlight utilization

Poles on One Side		Poles on Two Sides	
Width	Utilization	Width	Utilization
2 MH or less	50%	4 MH or less	50%
2–3 MH	40%	4–6 MH	40%
Over 3 MH	30%	Over 6 MH	30%

155,000 lumens. This lamp has a life of only 3000 hours, however, and is more suitable for short-burning-hour applications such as sports lighting. For the industrial area in the example, the 1000-watt HPS lamp rated at 140,000 lumens would be a better choice.

(Lumen ratings for lamps change frequently. The numbers in Table 2-1 are intended as estimates; always refer to the latest published lamp information for current ratings.)

Since 140,000 lumens falls between 100,000 and 150,000 lumens, the estimated four luminaires can probably be used.

In this case, there is a lamp size that satisfies the lumen requirements of the application with the estimated number of floodlights. This will not always occur. If the required level of illumination had been 108 lux instead of 54, for instance, the floodlight quantity would have had to be increased.

STEP 7

Check the photometric characteristics of the luminaire tentatively selected for the job. The floodlight that is finally selected will invariably be a compromise between the needs of the particular situation and the performance of the floodlight. The following are some guidelines to help in the selection.

Beam Spread The sample layout for the example (Fig. 10-1) shows that each floodlight must light a horizontal area which extends 45 degrees on either side of the direction in which the floodlight is aimed. The angle to each side of the actual aiming line is less than this, but the beam must extend beyond the edges of the area. This means that a floodlight with a beam spread of over 90 degrees (45 × 2) will be needed, with a National Electrical Manufacturers Association (NEMA) classification of 5 or 6 for the horizontal beam.

The vertical beam will have to cover the area from the aiming point back to the base of the pole. In the example, the floodlight is aimed at about 60 degrees, or just under 2 MH. Half the beam is 60 degrees; the total vertical beam spread will therefore be 120 degrees (60 × 2) and have a NEMA 6 rating. A floodlight with a NEMA 5 rating could also be considered, since two floodlights will be contributing to the area close to the pole and should provide enough illumination.

The asymmetric floodlight shown in Fig. 9-2 in the previous chapter has been tentatively chosen as the floodlight to use for the sample problem. It has a beam spread of 110 by 83 degrees, or NEMA 6 × 5, which would meet the needs of this particular application.

Maximum Candlepower at the Aiming Point In the sample problem, the floodlights will be mounted at 13.7 meters (45 feet) and aimed at a point approximately 2 MH, or 27.4 meters, from the pole base. The level of illumination at that point should not be too far above or below the average level, which in this case is initially 63.2 lux (5.7 footcandles).

Figure 9-6 in the previous chapter shows the relationship between the amount of candlepower required in order to produce 10.76 lux (1 footcandle) at the base of the pole and the amount required to produce the same light level at a distance of 2 MH. The formula used to make up the chart was

$$E = \frac{\text{cd} \times \cos^3 \theta}{\text{MH}^2}$$

where E = level of illumination
cd = candlepower in candelas

The cosine of the incident angle at the base of the pole is 1. To produce 63.2 lux at the base of the pole, then, would require

$$63.2 = \frac{\text{cd}}{13.7 \times 13.7}$$
$$= 63.2 \times 13.7 \times 13.7$$
$$= 11{,}862 \text{ candelas}$$

It requires 11.2 times as much candlepower, or 11.2 times as many candelas, to produce the same light level at a distance of 2 MH. Therefore, 11.2 × 11,862 = 132,854 candelas are needed.

The maximum candlepower of the tentative floodlight selection is 663 candelas per 1000 lamp lumens, as shown in the photometric data. Since the 1000-watt HPS lamp is rated at 140,000 lumens, the maximum candelas along the aiming line are (140,000/1000) × 663 = 92,820

This value is less than the required candela value that was calculated, but it is acceptable, since there will be some contribution from other floodlights. Also, this number is used only to establish the relative value of the maximum candlepower; it would be a coincidence if the selected floodlight matched the calculated value exactly. The maximum candlepower could range from one-third of 132,000 candelas, or 44,000 candelas, up to 132,000 candelas and still provide good uniformity. If it exceeded the candela value, though, the beam spread might be too narrow and the area might not be properly covered.

Utilization The utilization for the floodlight used in the sample problem was estimated to be 40 percent. This should be checked by referring to the candela and lumen distribution chart for the tentative floodlight selection, which in this case is given in Fig. 9-2 of the previous chapter. As noted, the figures along the left of the chart give a summation of the utilization of lamp lumens for the floodlight. In this case, the lower beam of the floodlight must cover 60 degrees from the aiming point to the base of the pole, because it is aimed at 60 degrees. The utilization figure for the block 60.0 degrees below the aiming point is 28 percent. The beam will also extend above the aiming point approximately 12 degrees, about to the edge of the area that is being lighted. (See Fig. 10-1.) The utilization for the second row of blocks in the upper beam, 12 degrees above the aiming point, is 15 percent. Since 28 + 15 = 43 percent utilization, which is close to the estimate, this floodlight will provide the needed utilization.

Uniformity

After determining the number of floodlights and making a preliminary choice of a specific floodlight to use, the designer must calculate the minimum level of illumination that can be tolerated. No lighting system is going to be absolutely uniform, no matter how carefully it is designed. So long as the maximum- to minimum-illumination ratios are less than 10:1, the system will appear uniform.

Unless a minimum level of illumination has been specified, the average-to-minimum ratio can be 3:1. In the example, the minimum level can be 63.2/3 = 21 lux (1.9 footcandles) initially.

The designer should study the photometric data for the floodlight being considered to see if these uniformity levels will be met. In the example, the floodlights will be aimed at a distance of just under 2 MH, which provides the best compromise between utilization and uniformity. The isoillumination diagram for this aiming point, as well as two associated tables of illumination levels, are given in Fig. 10-3.

The isoillumination diagram indicates that about 21.5 lux (2 footcandles) will be the level of illumination in the main part of the area at a 15.2-

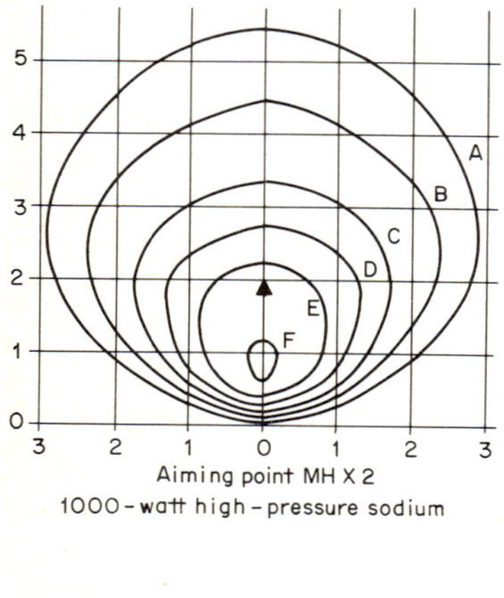

	Mounting Height 12.2 meters (40 feet) Lux	Footcandles	Mounting Height 15.2 meters (50 feet) Lux	Footcandles
A	1.7	0.16	1.1	0.1
B	3.3	0.31	2.2	0.2
C	8.4	0.78	5.4	0.5
D	17.2	1.6	10.8	1.0
E	33.4	3.1	21.5	2.0
F	83.9	7.8	54.0	5.0

Fig. 10-3 Isoillumination curve and table of illumination levels for the floodlight to be used for the sample problem. The aiming point will be 2 MH from the pole base.

meter (50-foot) mounting height and 33.4 lux (3.1 footcandles) at a 12.2-meter (40-foot) mounting height, as read from the table in Fig. 10-3. The value of line E at 13.7 meters (45 feet), the mounting height in the example, would be about 27 lux (2.5 footcandles). This is above the 21-lux minimum level which is needed for good uniformity.

However, if part of the area falls within line D, the lux level will be only 14, halfway between the 17.2 and 10.8 values on the chart. Before lighting this area, then, the designer should make four templates of the isoillumination diagram for the floodlight at this vertical aiming angle. These should be made to the scale of the working drawings, so that they can be manipulated on the drawings to show levels of illumination at various horizontal aiming angles. Remember that where lines cross, levels of illumination from these lines are added to yield the actual levels.

Determining the best horizontal aiming angle is often a matter of trial and error. In the example, there are two floodlights on a pole, and they will be aimed initially at around 45 degrees, which is midway in the area each floodlight is designed to cover. Some field adjustment can always be made if the minimum values are critical.

If isoillumination diagrams are not available, then a point-by-point method of checking illumination levels should be used.

Point-by-Point Calculations

Levels of illumination can be checked through various point-by-point methods. Figure 10-4 repeats the layout of the 54.9- by 54.9-meter area, which has now been divided into 6.1-meter (20-foot) square blocks. Levels of illumination can be measured at the center of any of these blocks to ensure that the average light level is being reached and uniformity is satisfactory. Good points to check are A, B, C, and D, especially if minimum light levels have been specified.

The most difficult part of point-by-point calculations is determining the distances and angles that the floodlight beam makes with the points selected, since these cannot be measured directly on a plan view of the lighted area. Figure 10-5 makes it possible for the designer to determine the level of illumination at any point, working only with the plan view and the photometric data for the floodlight to be used. The only dimensions required are the mounting height of the floodlight and the distance to the point where it is aimed.

The floodlight is located at the lower left corner of the chart. All measurements are made in relation to this point and must be dimensioned either in line with the aiming line or perpendicular to it. The aiming line is the left side of the chart.

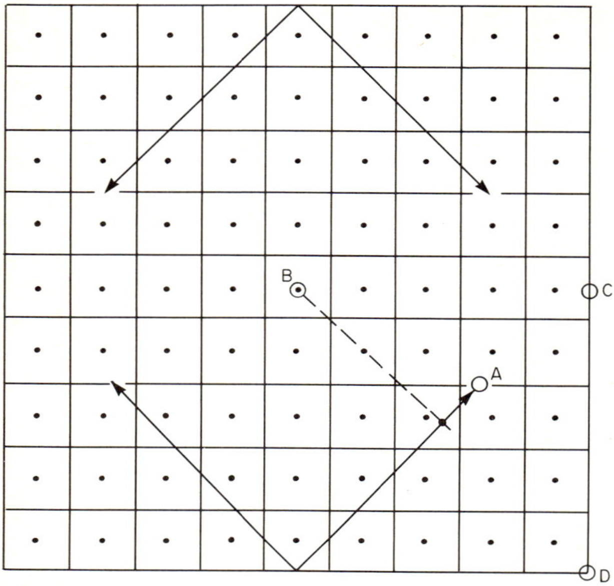

Fig. 10-4 Sample area divided into 6.1-meter (20-foot) blocks. Levels of illumination at points A, B, C, and D can be checked to ensure that the lighting system will fulfill the requirements of the particular job.

Floodlights are usually symmetrical left to right, so Fig. 10-5 can be used to check levels of illumination produced by either the left or right sides of the beam. Points to the left of the aiming line can be located on the chart in the same way as points to the right.

The grid line spacing is based on the mounting height of the floodlight. Along the left side of the chart are the vertical aiming angles associated with the grid. The aiming point in an actual situation can be determined either by this aiming angle or by the horizontal distance away from the floodlight location.

The horizontal aiming angle for any given point is indicated by the lines curving up from the base of the chart, off to the right. The value of this angle must be known so that candela information can be read from photometric data.

The circular lines are isoillumination lines, labeled with levels of illumination. The levels of illumination can be read as lux values; in this case, the chart would be based on a floodlight emitting 1000 candelas in all directions. Alternatively, the levels of illumination can be read as footcandles; in that case, the chart would be based on a floodlight emitting 10,000 candelas in all directions. The illumination levels are based on a mounting height of 10 meters (31.6 feet), which is the square root of 1000.

Figure 10-5 applies to any floodlight, because it is used in conjunction with specific photometric data. Table 10-2 reproduces the candela and lumen distribution chart for the asymmetric floodlight which will be used to light the 54.9- by 54.9-meter area in the example. Detailed information on reading this data is given in the previous chapter.

The designer can use the following procedures to check points that lie both on and off the aiming line of the floodlight, to see whether required levels of illumination are being met. This sample problem is based on lux and meters. The same procedures should be followed when using foot-candles and feet, except that allowances should be made for the variation in the candela values on which the chart is based.

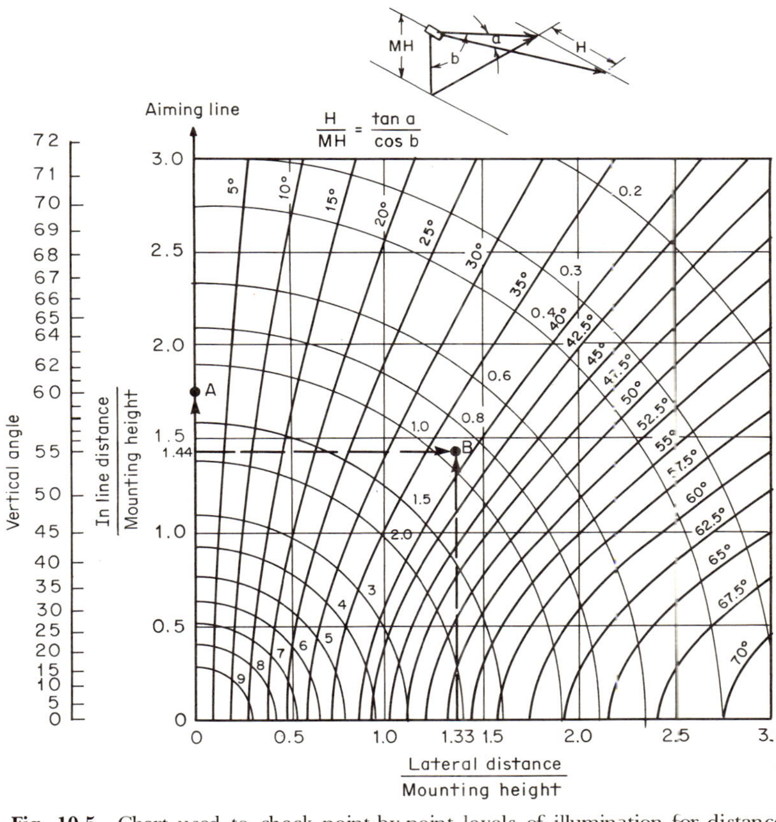

Fig. 10-5 Chart used to check point-by-point levels of illumination for distances up to 3 MHs. Points both on and off the aiming line can be checked. (In lux: illumination level per 1000 candelas, mounting height = 10 meters; in footcandles: illumination level per 10,000 candelas, mounting height = 31.6 feet.)

Table 10-2 Candela and lumen distribution—average left and right sides

Lamp Lumens	Lumens—Summation of One Side Only	85.0	75.0	65.0	55.0	45.0	35.0	25.0	15.0	5.0	0.0	(Lower Beam) Vertical Angles, degrees (Upper Beam)
		Horizontal Angles, degrees										
27	1.3	0 0.0	1 0.0	1 0.0	1 0.0	3 0.1	7 0.1	12 0.3	16 0.4	17 0.4	17	68.0
26	2.5	0 0.0	1 0.0	1 0.0	2 0.0	7 0.1	15 0.3	24 0.5	29 0.7	33 0.8	34	60.0
26	3.7	0 0.0	1 0.0	2 0.0	4 0.1	13 0.2	25 0.5	36 0.8	42 1.0	46 1.1	47	52.0
25	4.5	0 0.0	2 0.0	3 0.0	6 0.1	17 0.3	31 0.6	44 1.0	50 1.2	54 1.3	54	44.0
24	6.7	0 0.0	3 0.0	5 0.0	13 0.2	31 0.5	49 1.0	63 1.4	72 1.7	76 1.8	76	36.0
23	16.8	0 0.0	4 0.0	8 0.1	32 0.4	78 1.3	125 2.5	151 3.3	182 4.3	196 4.7	197	28.0
20	22.8	0 0.0	5 0.0	10 0.1	41 0.6	102 1.8	162 3.2	208 4.6	245 5.8	277 6.7	284	20.0
15	31.5	0 0.0	5 0.0	12 0.1	52 0.7	135 2.3	219 4.4	288 6.3	347 8.2	389 9.4	398	12.0
9	43.7	0 0.0	5 0.0	13 0.1	62 0.9	171 2.9	292 5.8	399 8.8	489 11.5	560 13.6	573	4.0
		0	5	13	64	181	315	441	553	645	663	0.0
9	44.1	0 0.0	5 0.0	14 0.1	63 0.9	173 3.0	295 5.9	403 8.9	495 11.6	566 13.7	578	4.0
15	32.6	0 0.0	5 0.0	11 0.1	53 0.7	139 2.4	226 4.5	298 6.6	361 8.5	404 9.8	412	12.0
20	25.0	0 0.0	4 0.0	10 0.1	45 0.6	109 1.9	176 3.5	229 5.1	269 6.3	307 7.4	311	20.0
24	18.5	0 0.0	3 0.0	7 0.1	33 0.5	84 1.5	135 2.7	173 3.8	198 4.7	220 5.3	228	28.0
27	13.4	0 0.0	3 0.0	5 0.1	25 0.3	61 1.1	102 2.0	127 2.8	143 3.4	153 3.7	156	36.0
28	3.9	0 0.0	2 0.0	4 0.0	9 0.1	16 0.3	26 0.5	30 0.7	48 1.1	49 1.2	56	44.0
28	3.0	0 0.0	2 0.0	3 0.0	6 0.1	10 0.2	16 0.3	23 0.5	37 0.9	43 1.0	49	52.0
28	1.7	0 0.0	1 0.0	3 0.0	5 0.1	7 0.1	10 0.2	11 0.3	19 0.4	25 0.6	25	60.0
29	1.1	0 0.0	1 0.0	2 0.0	3 0.0	5 0.1	7 0.1	9 0.2	9 0.2	16 0.4	17	68.0
	277	0.0	0.3	1.2	6.4	20.0	38.2	55.8	71.8	83.2		
		Lumens, Summation of Vertical Zones										
		28	28	28	28	27	25	21	15	8		

Note: Luminous intensity in candelas at centers of zones (upper nos.)
Luminous flux in lumens within zones (lower nos.)

CHECKING A POINT ON THE AIMING LINE

Step 1. Determine the point to be checked. For the example, point A in Fig. 10-4 will be checked first. This is the aiming point, and is at a vertical angle of 60 degrees from the floodlight. On Fig. 10-5, find the point along the aiming line (0 degrees) which is in line with the 60-degree point. It is about 1.75 on the scale of in-line distance expressed as a multiple of the mounting height.

Step 2. From the curved lines, find the illumination level at point A. The angle of 60 degrees is between the 1.0- and 1.5-lux points, and so a value of 1.25 lux can be estimated.

Step 3. From the photometric data, Table 10-2, find the number of candelas emitted at 0-0 degrees, the aiming point. In this case, the value is 663 candelas per 1000 lumens.

Step 4. From the manufacturer's data, find the initial lumens emitted by the lamp to be used. The value in this case is 140,000. Divide by 1000 to convert to lumens per 1000: 140,000/1000 = 140.

Step 5. Multiply the lumens per 1,000 of the lamp by the candelas per 1000 lumens at the 0-degree point: 140 × 663 = 92,820 candelas at 0 degrees.

Step 6. Divide this figure by 1000 to find the number of candelas per 1000 at that point, since the graph is based on 1000 candelas: 92,820/1000 = 92.82. (Note that if footcandle levels were being used the number from step 5 would be divided by 10,000, since the graph would be based on 10,000 candelas.)

Step 7. Multiply this figure by the illumination level at point A on the chart, which is 1.25 lux: 92.82 × 1.25 = 116. This is the lux level at the aiming point of 60 degrees, along the 0-degree line, for this luminaire and a mounting height of 10 meters.

Step 8. Since the mounting height in the example is 13.7 meters, not 10 meters, refer to Table 10-3 and multiply the lux level by the appropriate conversion factor. In this table, the mounting height of 13.7 meters lies between 12 meters and 14 meters. The conversion factor can be extrapolated; it is 0.54. 116 × 0.54 = 62.64 lux initial.

Step 9. Multiply by the light-loss factor to arrive at a maintained illumination level. Light-loss factor (LLF) = lamp lumen depreciation (LLD)

× luminaire dirt depreciation (LDD) = 0.9 × 0.95. 62.64 × 0.9 × 0.95 = 53.6 lux. This is almost exactly the desired maintained illumination level of 54 lux.

CHECKING A POINT OFF THE AIMING LINE

Step 1. Determine the point to be checked, in this case point B in Fig. 10-4. On the scale drawing for the job, draw a line from point B to the aiming line, so that a 90-degree angle is formed.

Step 2. Measure the horizontal distance from the base of the floodlight location to this intersection. For the example, this can be done with a ruler, since it is known that the drawing is divided into 6.1-meter blocks. It is 19.8 meters from the base of the pole to the point where the line from B intersects the aiming line.

Step 3. Measure the horizontal distance from this intersection to point B. In the example, it is 18.3 meters.

Step 4. Divide each value by the mounting height, which is 13.7 meters, in order to find the values to use on the coordinates of Fig. 10-5: in-line distance/mounting height = 19.8/13.7 = 1.44; lateral distance/mounting height = 18.3/13.7 = 1.33.

Step 5. Locate point B on Fig. 10-5 by drawing a line from the value of 1.44 on the left ordinate to meet a perpendicular line drawn from the value 1.33 on the lower ordinate.

Step 6. On the left side of the chart find the vertical angle from the pole base by extending the line drawn to point B. In this case it is about 55 degrees, or 5 degrees below the aiming point of 60 degrees. The horizontal angle is given on the curved lines. Point B is between 35 and 40 degrees, or at about 37 degrees.

Table 10-3 Mounting height correction factors

Multipliers for Other Mounting Heights, Meters													
4	6	8	10	12	14	16	18	20	22	24	26	28	30
6.25	2.78	1.56	1.0	0.69	0.51	0.39	0.31	0.25	0.21	0.17	0.15	0.13	0.11
Multipliers for Other Mounting Heights, Feet													
10	15	20	25	30	35	40	45	50	60	70	80	90	100
10	4.4	2.5	1.6	1.1	0.816	0.625	0.493	0.4	0.28	0.2	0.156	0.12	0.1

Multipliers for other mounting heights:
when using lux: $100/MH^2$
when using footcandles: $1000/MH^2$

Step 7. Find the lux level at point B. It is between the lines for 0.8 and 1.0 lux, and is therefore 0.9 lux.

Step 8. Refer to the candela and lumen distribution chart, Table 10-2, to find the number of candelas per 1000 lumens at point B. The first row of blocks in the lower beam includes the 5-degree vertical angle. Follow this first row to the horizontal angle of 37 degrees. At the 35-degree point, the luminaire emits 295 candelas per 1000 lumens, and at 45 degrees, 173 candelas. For 37 degrees, then, estimate a value of 270 candelas per 1000 lumens.

Step 9. Find the lumens per 1000 for this floodlight: 140,000/1000 = 140.

Step 10. Multiply the lumens per 1000 of the lamp by the candelas per 1000 at point B: 140 × 270 = 37,800 candelas at point B.

Step 11. Divide this figure by 1000 to find the candelas per 1000: 37,800/1000 = 37.8.

Step 12. Multiply this figure by the lux level at point B, which is 0.9: 37.8 × 0.9 = 34 lux at point B at a mounting height of 10 meters.

Step 13. Refer to Table 10-3 to find the conversion factor for a mounting height of 13.7 meters: 34 × 0.54 = 18.36 lux initially for each floodlight.

Step 14. Since four floodlights will be contributing equally to the level of illumination at point B, multiply the 18.36 value by 4 to arrive at the actual lux level at that point: 4 × 18.36 = 73.44 lux.

Step 15. Multiply by the light-loss factor: 73.44 × 0.9 × 0.95 = 62.8 lux. This is the maintained level of illumination at point B. It is well above the average level of 54 lux.

Point-by-Point Calculations for Large Areas

Figure 10-6 is used in exactly the same manner as Fig. 10-5. The multipliers for other mounting heights included in Table 10-3 apply to Fig. 10-6 as well. Fig. 10-6 is intended for use when floodlights have been aimed at very high angles. For areas where floodlights have been aimed at between 0 and 3 MH, Fig. 10-5 should be used to ensure more accurate values.

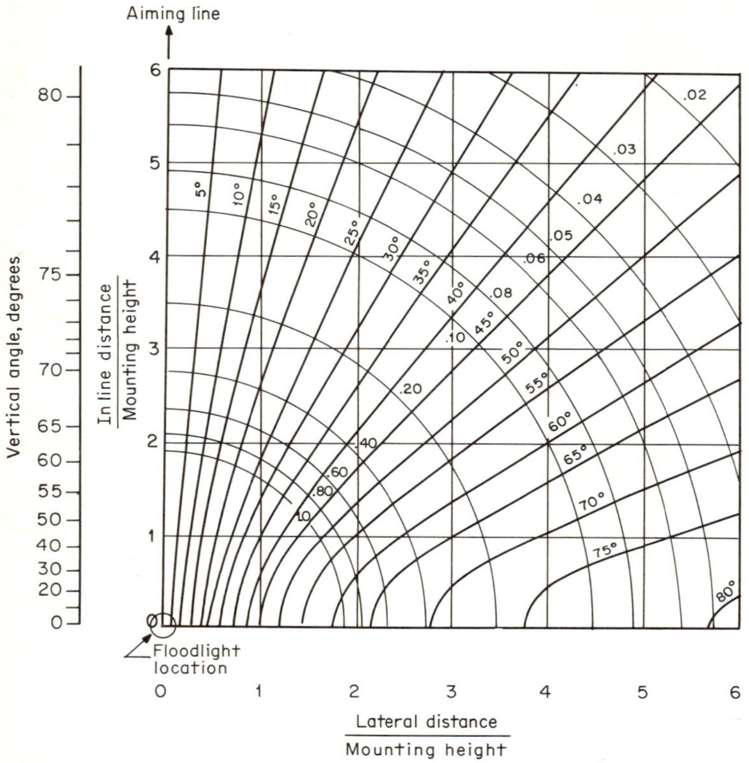

Fig. 10-6 Chart similar to Fig. 10-5, but designed to be used to determine illumination levels as far as 6 MH from the pole base. (In footcandles: illumination level per 10,000 candelas, 31.6-foot mounting height; in lux: illumination levels per 1000 candelas, 10-meter mounting height.)

For complicated layouts, levels of illumination for many points in the lighted area can be calculated by using the appropriate chart. Illumination values in each block in the photometric data can be tabulated for a specific aiming angle and mounting height. These can be noted on the scale drawing of the area. Points of equal illumination value can then be connected to produce isoillumination curves.

Point-by-Point Calculations by Computers

Some manufacturers provide computer services which will calculate levels of illumination at any point in an area lighted by one floodlight or an array of floodlights. These computer programs can calculate maximum, minimum, and average levels of illumination, on horizontal or vertical planes. There are also methods of calculating illumination values as seen by a television camera. This is useful for designing surveillance systems.

The use of computers is the most accurate way to calculate the actual performance of a floodlighting system, but this added accuracy is not necessary for small jobs or uncomplicated layouts. By following the guidelines in this chapter and the preceding one, the designer will be able to plan a satisfactory floodlighting system for almost any type of outdoor application.

Bibliography

Hopkinson, R. G., and J. B. Collins: *The Ergonomics of Lighting,* Macdonald & Co., Publishers, Ltd., London, 1970, pp. 17–27.

IES Lighting Handbook, 5th ed., Illuminating Engineering Society of North America, New York, 1972, pp. 9-90, 9-91.

Musick, J. E.: "Graphic Aids for Floodlighting Design," *Illuminating Engineering,* October 1969, pp. 625–630.

11

Photometric Performance of Roadway, Area, and High-Mast Luminaires

Outdoor areas can be lighted with roadway, area, or high-mast luminaires, as well as with floodlights. The major difference among these systems is that the vertical aiming is already fixed in roadway, high-mast, and area luminaires, whereas floodlights must be aimed.

The photometric performance of floodlights is explained in Chap. 9, and the actual design procedures in Chap. 10. This chapter will describe how to use high-intensity-discharge roadway, area, and high-mast photometric data and will provide other information that the designer needs before choosing a lighting system. The next chapter will explain design procedures for outdoor area lighting, parking lots, and roadway lighting, using these luminaires.

Types of Systems

There are three types of outdoor luminaires with fixed vertical aiming: roadway, area, and high-mast. These luminaires are designed to be mounted in a position parallel to the ground, so that their light pattern falls straight below the luminaire. The orientation of roadway and area luminaires is fixed with the position of the mounting bracket. High-mast luminaires are also fixed into position, but the optical assembly within the luminaire can be rotated to change the orientation of the light pattern. The pattern still falls directly below the luminaire, but it can be oriented in any direction in relation to the pole.

The designation *area luminaire* applies to any luminaire with fixed aiming that is not a roadway luminaire. Roadway luminaires have a definite beam spread which is designed to light roadways. Area luminaires are usually used to light parking lots.

Generally, roadway and area luminaires are used at mounting heights where they can be serviced from the ground. High-mast luminaires are used at higher mounting heights, on poles equipped with a lowering device. This is not always the case, however. There is no set height that qualifies a pole as a high mast; the term applies to a pole of any height that is equipped with a lowering device for servicing the luminaires.

Roadway, area, and high-mast luminaires are designed to be mounted individually or around a pole. They light in all directions away from the pole, and so no great amount of light is projected in any one direction. Consequently these luminaires do not have high maximum candlepower, as some floodlights do.

Presentation of Photometric Data

Photometric data for roadway, area, and high-mast luminaires is presented in the same manner. All data is presented as if a roadway were the area to be lighted, but the information can be used to design an effective lighting system for any kind of outdoor area.

At the time of this writing, photometric data is still presented in footcandles and feet. In some cases the authors have been able to present data in lux and meters as well as footcandles and feet, but this has not always been practical.

All photometric data is based on the orientation of the luminaire as shown in Fig. 11-1. In the diagram, 0 lateral degrees is straight in front of the luminaire, and 0 vertical degrees (the nadir) is straight beneath it. Light in the region of 90-0-270 lateral degrees is said to be emitted on the *street side,* since that light would normally be directed over the street or roadway. Light between 90-180-270 degrees is said to be directed toward the *house* or *curb side,* since that region would normally be behind a road-

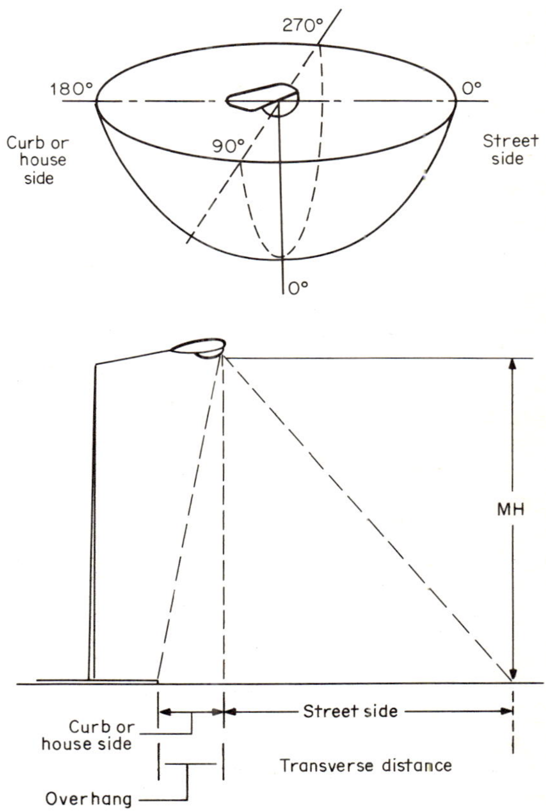

Fig. 11-1 Orientation of a luminaire used as the basis for roadway, area, and high-mast photometric data.

way luminaire, toward the houses. If the luminaire overhangs a street, then some of the curb- or house-side light will be directed onto the roadway, as shown.

Upward light is that which is directed above 90 degrees vertical. *Downward* light is directed below 90 degrees vertical, which in effect is the equator of the sphere in Fig. 11-1.

Classification of Luminaires

The *American National Standard Practice for Roadway Lighting,* sponsored and published by the Illuminating Engineering Society of North America (IES), defines the criteria for the classification of roadway, area, and high-mast luminaires. The classifications used in most photometric data are:

1. Vertical light distribution
2. Lateral light distribution
3. Control of light distribution above maximum candlepower

All classifications used in photometric data are based on lighting a strip running perpendicular to the crossarm of the luminaire, as a roadway would. Photometric data applies to a continuous horizontal section with two boundaries, which are referred to as the *near* and *far curb lines. Transverse* areas or lines are those which run perpendicular to the curb of the roadway, and *longitudinal* strips or lines run parallel to the curb.

VERTICAL LIGHT DISTRIBUTION

Luminaires are rated as short, medium, or long in their vertical light distribution. These ratings are based on where the maximum candlepower from the luminaire falls on a roadway surface in relation to transverse roadway lines, as shown in Fig. 11-2. The distances marked by the transverse roadway lines (TRL) and the longitudinal roadway lines (LRL) are given in the diagram as multiples of luminaire mounting height (MH).

The transverse lines divide the roadway into zones which extend all the way across the roadway. The first zone is bounded by the transverse roadway lines drawn at distances of 1 and 2.25 MH from the luminaire. This is the *S zone,* and if the maximum candlepower from the luminaire falls within this zone, the distribution is rated as short.

The next zone is bounded by the lines marking distances of 2.25 and 3.75 MH. The distribution of a luminaire is rated as medium if the maximum candlepower falls within this second zone, the *M zone.* The luminaire depicted in Fig. 11-2 has a medium distribution.

The last zone is the *L zone.* It extends from 3.75 to 6.0 MH. If the maximum candlepower from a luminaire falls within this zone, the distribution is rated as long.

These ratings are determined by the vertical angle of maximum candlepower of the luminaire. Roadway and high-mast luminaires are aimed straight down, but they are designed to deliver their maximum candlepower at angles up to 80 degrees. The ratings of vertical light distribution are given below, along with the vertical angle of maximum candlepower which each rating represents:

1. Short distribution: 1.0-MH TRL to 2.25-MH TRL: approximately 45 to 66 degrees vertical angle.

2. Medium distribution: 2.25-MH TRL to 3.75-MH TRL; approximately 66 to 75 degrees vertical angle.

3. Long distribution: 3.75-MH TRL to 6.0-MH TRL; approximately 75 to 80 degrees vertical angle.

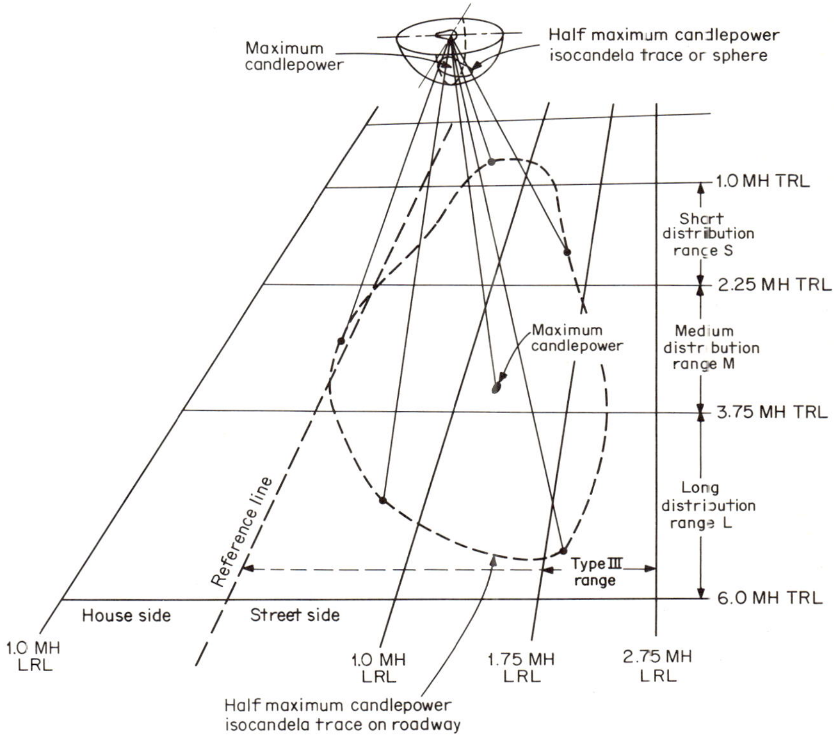

Fig. 11-2 Classifications of vertical light distribution are based on where the maximum candlepower from a luminaire falls in relation to transverse roadway lines. Ratings for lateral light distribution depend on where the one-half maximum candlepower isocandela trace falls. (This drawing is reprinted with permission from *American National Standard Practice for Roadway Lighting,* ANSI/IES RP-8, 1977.)

LATERAL LIGHT DISTRIBUTION

The lateral light distribution of a luminaire is indicated on manufacturer's photometric data by the classification of the luminaire as type I, II, III, IV, or V. This rating gives a rough indication of the beam width of the luminaire and indicates how far across the road or area light will be distributed.

The type classification should never be used alone to determine which luminaire to use, but the width of the beam should be selected to come as close as possible to fitting the width of the area to be lighted.

Changes in type from I to IV can be accomplished by changing the lamp within the reflector or by changing the reflector, refractor, or both. The method to use is decided by the luminaire manufacturer.

The type of a luminaire is determined by the position of an isocandela trace connecting all the points in the lighted area which will receive a value of half the maximum candlepower of the luminaire. Such a trace is shown in Fig. 11-2. The relation of the one-half maximum candlepower isocandela trace to the longitudinal roadway lines determines the type except for type V, which has a circular distribution.

Figure 11-2 has been divided into sections bounded by longitudinal roadway lines. Indicated on the diagram is a reference line drawn directly in line with the mounting position of the luminaire. Distances from this line are given as multiples of luminaire mounting height.

Following are the type classifications for lateral light distribution:

Type I: One-half maximum candlepower isocandela trace enters the area on both sides of the reference line and remains within the area bounded by 1.0-MH LRL on both the house and street sides within the S, M, or L zone where the maximum candlepower falls. Luminaires with a type I classification are usually mounted at or near the center of the roadway or area. Types II, III, and IV are generally mounted at the side of the roadway or area.

Type II: One-half maximum candlepower isocandela trace does not cross the 1.75-MH LRL on the street side in the zone where the maximum candlepower falls. The line may or may not cross the reference line, but should stay close to it.

Type III: One-half maximum candlepower isocandela trace crosses the 1.75-MH LRL but does not cross the 2.75-MH LRL on the street side in the zone where the maximum candlepower falls. The line may or may not cross the reference line. The luminaire depicted in Fig. 11-2 is type III.

Type IV: One-half maximum candlepower isocandela trace crosses the 2.75-MH LRL in the zone where the maximum candlepower falls. The line may or may not cross the reference line.

Type V: Candelas are symmetrically distributed at all lateral angles around the luminaire.

CONTROL OF LIGHT DISTRIBUTION ABOVE MAXIMUM CANDLEPOWER

Classifications concerning control of light distribution above maximum candlepower are also given by manufacturers in photometric data. The classifications indicate the candelas emitted from the luminaire at high angles. The IES classifications are as follows:

Cutoff: Not more than 25 candelas per 1000 lamp lumens above an angle of 90 degrees above the nadir, and not more than 100 candelas per 1000 lumens above 80 degrees above the nadir. These criteria apply at any lateral angle around the luminaire, and both must be met in order for a luminaire to be classified as cutoff.

Semicutoff: Not more than 50 candelas per 1000 lamp lumens above an angle of 90 degrees, and not more than 200 candelas per 1000 lamp lumens above 80 degrees, at any lateral angle.

Noncutoff: No candela limitations.

These ratings are important, because the more candelas that are emitted at high angles, the brighter the luminaire will appear. In some cases, it is desirable to have all or almost all of the light output stop at 90 degrees, in order to reduce the brightness of the luminaire to near zero beyond a distance of 10 MH. This is sometimes accomplished by using a flat glass lens instead of a protruding refractor in the luminaire.

Utilization Data

Utilization data shows the light output of a luminaire. It indicates what portion of the light falls in front of the luminaire's location above the roadway (that is, on the street side) and what portion falls behind that location (that is, on the curb side). The data is presented in the form of a graph which gives the percentage of lamp lumens reaching the roadway at various ratios of transverse distance to mounting height. The data applies to streets of any width, if the width is given as a function of mounting height.

Figure 11-3 give photometric data for a roadway luminaire rated medium, semicutoff, type III. Included is the chart labeled Utilization Curve. The dashed line indicates the utilized lumens, or coefficient of utilization (CU) for the street side, and the solid line the CU for the curb

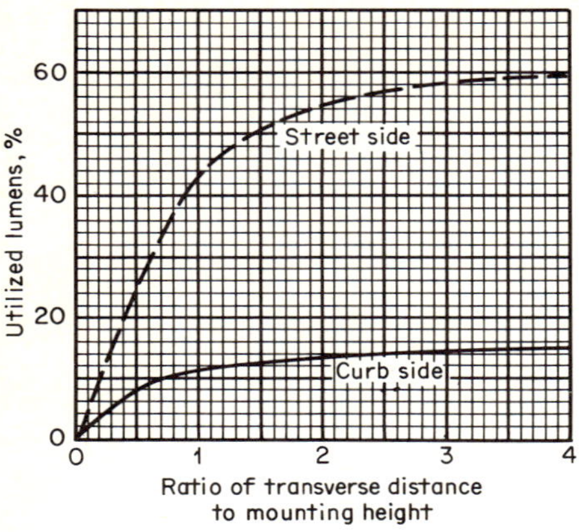

Utilization curve

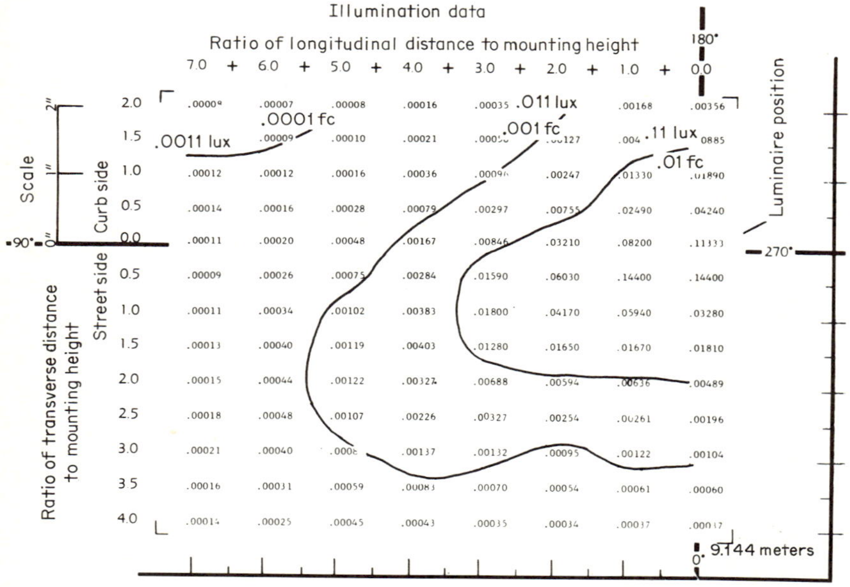

Illumination data is based on a luminaire mounting height of 9.144 meters (30 feet). For other mounting heights multiply the values of illumination shown by the factors in the following table:

MOUNTING HEIGHT						
Meters	7.62	9.144	10.668	12.19	13.7	15.24
Feet	25	30	35	40	45	50
FACTOR	1.44	1	0.74	0.56	0.44	0.36

Fig. 11-3 Photometric data for a medium, semicutoff, type III high-pressure sodium roadway luminaire.

Light-flux values

	Lumens	Percent of Lamp
Downward, street side	600	60
Upward, street side	10	1
Downward, curb side	150	15
Upward, curb side	10	1
Total	770	77

General Information:
Test number: 74-0525
Test distance, meters: 7
Test lumens: 1000
If the rating of the lamp used differs from the test rating of 1000 lumens, multiply all lumen, candela (if shown), and footcandle values by this ratio:

$$\text{Ratio} = \frac{\text{actual lamp lumens}}{\text{test lumens}}$$

Maximum candlepower = 750
Maximum cone = 70°
Maximum vertical plane = 72.5°/287.5°
Maximum candlepower at 90° = 22
Maximum candlepower at 80° = 174
Nadir footcandles = .113
Nadir candlepower = 102
Photometric test in accordance with IES guide.

side. If the luminaire is mounted directly over the curb, then only the street-side utilization applies. If the luminaire is mounted out over the street, then some of the curb-side lumens will also fall on the roadway.

For a type V luminaire, the street-side and curb-side utilization curves are the same, since light is distributed in a circular pattern. All other types of luminaires have asymmetric beam patterns.

KNEE OF THE CURVE

A straight utilization curve would indicate uniform distribution of lamp lumens across the road. Most curves taper off at a certain transverse line, and this bend in the curve indicates that light output is dropping. The section of the curve where the line flattens out is sometimes referred to as the *knee of the curve*. In Fig. 11-3, for example, the knee of the curve occurs between 1 and 2 MH.

The knee of the curve can help the designer to determine the approximate width of the area in front of the luminaire which can be effectively lighted; it does not help in the determination of horizontal spacing or uniformity.

MOUNTING HEIGHT

Utilization data can help the designer to choose a mounting height for a given situation. If the knee of the curve for the luminaire to be used occurs at 1 or 2 MH, for instance, the pole height can be approximately half the width of the street or area. If the street is 20 meters wide, a

10-meter pole can probably be used, since the luminaire can efficiently project light from the luminaire to a distance of 2 MH.

Sometimes higher mounting heights are necessary, however, in order to project light further down the street and thus increase luminaire spacing. This is particularly true of narrow streets or areas. The designer may also find it necessary to increase mounting height in order to meet uniformity requirements.

CORRELATION OF UTILIZATION CURVES AND TYPE RATINGS

There is a rough correlation between utilization curves and type ratings. For a type I luminaire the street-side and curb-side utilization curves are usually close together, and they get farther and farther apart as the type number gets higher. For this reason the designer would be less likely to overhang a type IV luminaire than a type I. Types II, III, and IV are usually mounted over or near the edge of the roadway. Types I and V are often mounted near the center of the area to be lighted.

Before choosing which type of luminaire to use, the designer should study the family of utilization curves in a full range of light distributions. The luminaire with the best combination of utilization and uniformity can then be chosen.

Illumination Data

Included in the photometric data for any luminaire is a chart labeled Illumination Data, as shown in Fig. 11-3. The numbers in the chart stand for footcandles; each number can be converted to lux by multiplying by 10.76.

This data is presented as if the area to be lighted were a street, but it applies equally well to any area. The luminaire is mounted in the spot marked "luminaire position." Data is given for only one side of the luminaire beam, since values are almost the same from left to right and are averaged to yield illumination data for one side.

The coordinates are based on ratios of distance to mounting height. The data in this case applies to a 9.1-meter (30-foot) mounting height but can be converted to other mounting heights by using the table included with the data.

The numbers on the chart are levels of illumination per 1000 lamp lumens at various points within the lighted area. The isoillumination lines connect the points of equal illumination, given in this case first in lux and then in footcandles. The data can be applied directly to the area to be lighted, since roadway, area, and high-mast luminaires have fixed aiming.

Sometimes the dimensions of the lighted area can be drawn directly on the illumination data to show levels of illumination at various points. These values can then be adjusted for mounting heights other than 9.1 meters and for either lux or footcandles by using the correction factors included with the data. Values are expressed per 1000 initial lamp lumens and should be adjusted to fit the actual lumen rating of the lamp to be used.

It is helpful in some cases to redraw illumination data to the same scale as that used in the drawing of the proposed lighting layout. By superimposing this diagram on the layout, the designer can study the distribution of light. Two or more diagrams can be laid on a drawing, as if they were light coming from two or more luminaires, and tried at various distances from each other. Where isoillumination lines cross, the values of individual lines are added together to yield actual levels of illumination. This gives a good indication of how uniform a lighting job will be, and also helps the designer to determine final spacing.

Spacing and Uniformity

The formula to determine luminaire spacing is

$$S = \frac{\text{LL} \times \text{CU} \times \text{LLF}}{E \times W}$$

where S = spacing
LL = initial lamp lumens
CU = coefficient of utilization
LLF = light-loss factor
E = level of illumination
W = width of road or area

If everything except spacing and initial lamp lumens is determined, or constant, then spacing becomes a function of lamp lumens.

For any one source type, lamp lumens can be increased simply by using a higher-wattage lamp. For instance, high-pressure sodium (HPS) sources are available in 50-, 70-, 100-, 150-, 200-, 250-, 310-, 400-, and 1000-watt sizes, each with a different lumen rating. Theoretically, then, the highest-wattage source could be used at a very wide spacing. In practice, however, this does not always provide a satisfactory lighting system. Other factors must be considered, particularly the effect on uniformity of a change in spacing and wattage.

The final decision on spacing should always be limited by the uniformity desired, which should be kept close to an average-to-minimum ratio of 3:1. Uniformity can be determined with the help of isoillumination diagrams, as explained above, or by using computer design programs, which are available from some manufacturers. The use of computers is the most accurate way to tie together the variables of mounting height, overhang, lamp size, spacing, and distribution. There is no other quick way of determining which combination is best.

Most manufacturers publish tables that suggest which lamp types to use for various illumination levels and road widths in order to produce 3:1 average-to-minimum uniformity. Recommended mounting heights and spacings are also given. These tables are helpful in design.

Placement of Poles along a Roadway

There are three ways in which poles are generally placed along a roadway: on one side, opposite each other, or staggered. Staggered or opposite arrangements are generally best for roadway lighting. Placing poles on both sides of a roadway does add to wiring costs, however, and a one-side arrangement is less expensive in that respect.

For a two-lane, lightly traveled roadway within an industrial complex, a row of luminaires on only one side of the road would probably provide adequate coverage. The beam pattern of the luminaire should be chosen to fit the width of the roadway as closely as possible, so that uniformity can be kept within reasonable ratios.

Light-Flux Values

Photometric data usually includes a chart of light-flux values, as shown in Fig. 11-3. This chart divides the distribution of lamp lumens from the luminaire into quadrants: downward or below a 90-degree vertical angle on the street side; upward or above a 90-degree vertical angle on the street side; downward on the curb side; and upward on the curb side. The data also includes the total lumens emitted by the luminaire. In this example, all figures are expressed per 1000 lamp lumens. For high-intensity discharge (HID) and filament lamps, data is also given for the percentage of rated lamp lumens that each lumen value represents. These numbers are used to show the efficiency of the luminaire.

General Information

The section of the photometric data labeled *general information* (Fig. 11-3) includes such things as the test number, the test distance, and the constant number of lumens that were used in the testing (in this case, 1000). If the rating of the lamp to be used differs from the test rating, the conversion formula is given as a ratio of actual lamp lumens to test lumens

The rest of the data in this section summarizes the photometric data which is used to determine the luminaire type. The *maximum cone* of the luminaire is the vertical angle at which the maximum candlepower occurs. The maximum candlepower from the luminaire used in the example is delivered at a vertical angle of 70 degrees from the nadir. The angle of maximum candlepower for area luminaires or type V roadway or high-mast luminaires occurs in a circle, so that it is the same regardless of the direction from the luminaire. Type I, II, III, and IV roadway and high-mast luminaires emit a beam shaped like a double lobe; the maximum cone occurs in two vertical planes. These vertical planes are measured from the 0-degree point, which is directly in front of the luminaire. The lateral angles are numbered in a circle from that point clockwise around the luminaire.

In Fig. 11-3 it can be seen that the maximum vertical plane for this luminaire occurs at lateral angles of 72.5 and 287.5 degrees. This means that the maximum candlepower from this luminaire is delivered at these two lateral angles, as shown in Fig. 11-4.

The maximum candlepower at 90 degrees above the nadir for this luminaire is 22 candelas per 1000 lamp lumens, and the maximum candlepower at 80 degrees is 174 per 1000 lamp lumens. The 90-degree rating would be sufficient for the luminaire to be rated as cutoff, but the 80-degree rating is not, and so the rating is semicutoff.

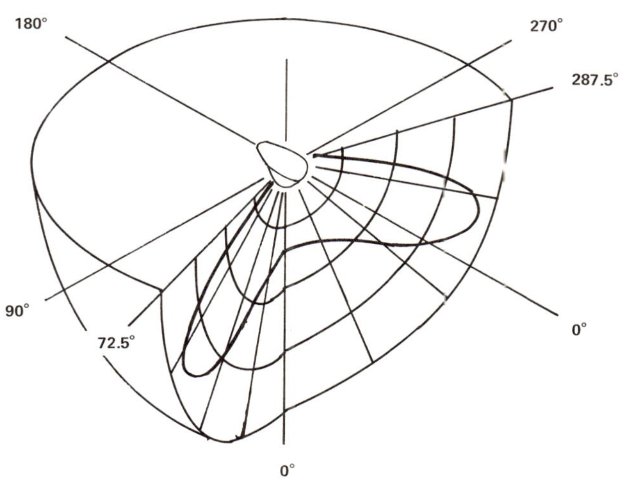

Fig. 11-4 Drawing showing the two angles at which maximum candlepower is emitted from the roadway luminaire depicted in Fig. 11-3. All type I, II, III, and IV roadway and high-mast luminaires emit a beam shaped like a double lobe.

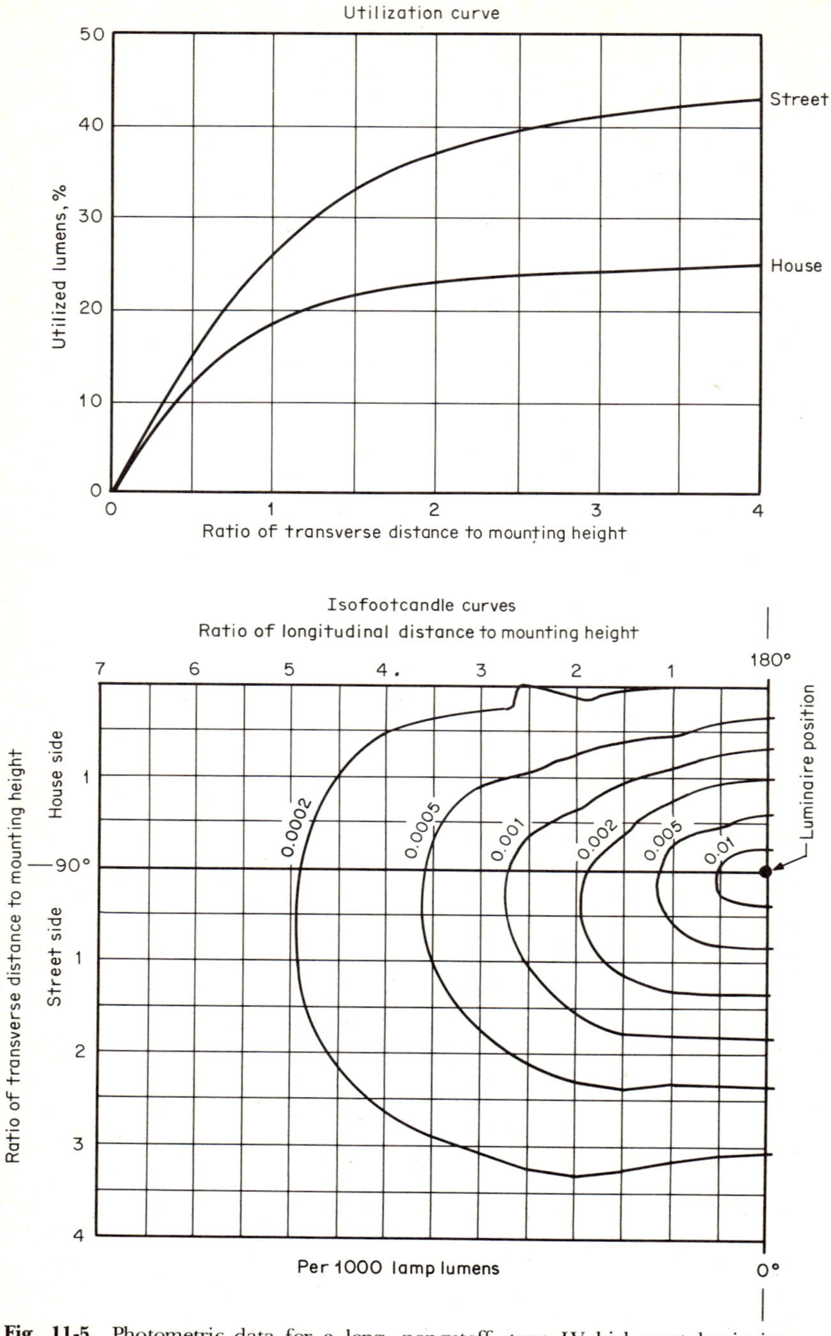

Fig. 11-5 Photometric data for a long, noncutoff, type IV high-mast luminaire. (1 footcandle = 10.76 lux.)

Nadir footcandles is the footcandle level directly below the luminaire, and *nadir candlepower* is the number of candelas directed straight down.

High-Mast Photometric Data

High-mast photometric data is presented in almost exactly the same manner as that for roadway and area luminaires. For convenience, however, the illumination data is usually based on a 30.5-meter (100-foot) mounting height. The general information given may also be different or arranged in some other form than that given for roadway luminaires. Figure 11-5 gives photometric data for a typical high-mast luminaire rated as long, noncutoff, type IV.

UTILIZATION CURVES

Figure 11-5 includes the utilization curve for the luminaire. This particular luminaire is asymmetric, so two curves are given. Symmetrical luminaires have only one utilization curve that can be used for either the street side or the curb side. Values can be doubled to yield total utilization. Utilization curves for high-mast luminaires are read in the same way as utilization curves for roadway luminaires.

ISOILLUMINATION CURVES

The isoillumination curves for this high-mast luminaire are shown in Fig. 11-5. In this case, curves are labeled *isofootcandle curves,* since data is given in footcandles. They serve the same purpose as the isoillumination curves for roadway and area luminaires. The luminaire position is the same, and the coordinates are based on their relation to mounting height. The isoillumination lines connect the points in the lighted area that will have the same level of illumination.

These lines can be used to determine levels of illumination at various points, as well as uniformity. This can be done by working directly with the isoillumination curves or by redrawing them as templates to be used on working drawings. Where lines cross, values are added to yield actual

Mounting Height Correction Factors for Other Than 100 feet
Use Table below or Factor = $10{,}000/(\text{actual MH})^2$

Mounting height, feet	80	90	100	110	120	130	140
Factor	1.56	1.23	1.00	0.83	0.69	0.59	0.51

Max. cone	60
Max. candela	212
Cone	*Candela*
0°	70
5°	69
15°	61
25°	69
35°	90
45°	129
55°	191
60°	212
65°	186
70°	148
75°	102
80°	63
85°	26
90°	10

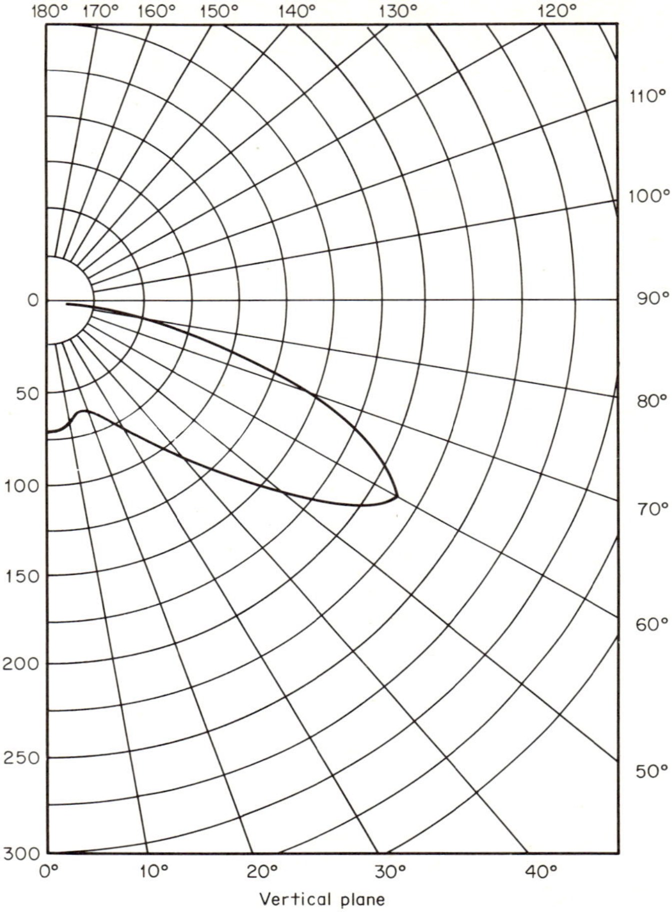

Fig. 11-6 Drawing of the vertical plane of a symmetrical high-mast luminaire. The table gives the exact candela values for each vertical angle.

light levels. Levels of illumination are based on 1000 lamp lumens and must be adjusted for the initial lumens of the lamp that is to be used.

In this case, isoillumination data is based on a mounting height of 30.5 meters (100 feet). Figure 11-5 gives the correction factors for other mounting heights.

ADDITIONAL DATA FOR SYMMETRICAL LUMINAIRES

Some manufacturers publish additional photometric data for symmetrical high-mast luminaires. Figure 11-6 shows the vertical plane for the luminaire used in the previous example. For a symmetrical high-mast luminaire, this graph indicates the candela distribution at every plane from the base of the pole. The left side of the graph is the number of candelas at each vertical angle on the radial coordinates.

These graphs are not given for asymmetric high-mast luminaires, because the values would be different for each lateral angle.

Along with the graph of the vertical plane there is usually a table giving the exact candela value for each vertical angle (see Fig. 11-6). The table shows that the maximum cone for this luminaire occurs at 60 degrees. Since this is a symmetrical luminaire, the maximum cone is at 60 degrees at all lateral angles.

With an asymmetric high-mast luminaire, the maximum cone occurs at two angles. This is indicated in photometric data by the lateral angles labeled *max vertical plane*. This is the same system as that used in roadway and area photometric data.

High-Mast Orientation

The major difference between high-mast and roadway and area luminaires is that high-mast luminaires can be oriented by rotating the optical assembly. (There is no orientation for a type V luminaire, since the beam pattern is symmetrical; types I, II, III, and IV luminaires can be oriented.) There are several ways of doing this.

1. Mount all the luminaires in a circle on the pole, but orient the optical assemblies in the same way, so that they are all parallel to one another and facing the same way. This is often done when the pole is on or near the perimeter of the lighted area.

To arrive at actual levels of illumination, multiply values for one luminaire by the number of luminaires that are used. This gives illumination levels for the whole system. The distance between luminaires is so slight in comparison to the mounting height that all the luminaires can be considered as a unit.

2. Mount luminaires back to back, with one half oriented in one direction and one half oriented in the opposite direction. The pattern of light produced will then be the same on either side of the pole. This is done when poles are in the center of an area. If a type IV luminaire is used, the pattern of light will be rectangular.

It is more difficult to calculate levels of illumination when luminaires are oriented in this manner, because the contribution of each must be calculated separately, overlapped with data from other luminaires, and the results must be added. This can be done with templates or with help from the manufacturer.

3. Mount luminaires in a radial pattern with each optical assembly oriented away from the pole. This will produce a circular pattern of light. It is easiest to do this with symmetrical, type V luminaires, because expected levels of illumination for one luminaire can then be multiplied by the number of luminaires to yield system values. Sometimes, however, a type V luminaire does not have as high a maximum candlepower or maximum cone as luminaires of other types. Then a cluster of asymmetric luminaires might be a better choice, because they would cover a larger area. The best choice can be determined only by comparing the photometric data for all type classifications for the luminaire to be used.

If asymmetric luminaires are used, a circular pattern of light can be produced with this orientation by using six or more luminaires. If four are used, the light pattern will be closer to a square.

4. Mount luminaires in any combination of beam spreads that fit the shape of the area. This is the most difficult pattern, since photometric data for each luminaire must be fitted into the lighted area, and the results must be added to yield actual levels of illumination. This can be done fairly laboriously with templates, or the advice of the luminaire manufacturer can be followed. The most accurate method is to use the computer service offered by some manufacturers, which will calculate the levels of illumination produced by any array of luminaires. These programs will also give average light levels and uniformity ratios.

Choosing a Pole Height

Factors for the designer to consider in choosing pole height are utilization, uniformity, community regulations, shadows, and cost. Pole costs increase rapidly with higher mounting heights, but the area covered from one pole goes up as the ratio of the square of the mounting height for the shorter pole to the square of the mounting height of the taller pole. A system mounted on a 12.2-meter (40-foot) standard pole will light approximately 3950 square meters (1 acre). A 45.7-meter (150-foot) high-mast system can light 39,500 square meters (10 acres).

The least expensive lighting system can be obtained by using the tallest high-mast pole, since pole costs do not rise in proportion to the size of the area that can be lighted. Adding a lowering device increases the cost of the pole, but this can be offset by using fewer poles and higher-wattage, more efficient lamps, such as 1000-watt HPS.

High-mast systems less than 30.5-meters (100-feet) high can be higher in initial cost than systems using standard poles. High masts may still be the best choice if their use would reduce maintenance costs or if it would be advantageous to have as few poles as possible in the area. (Fewer poles mean fewer shadows, since each luminaire location will cast its own shadow.) It is less expensive to add luminaires to a pole and/or increase pole height than it is to add poles.

The most efficient system will be the one using the least number of poles and the highest-wattage sources, but the final choice is limited by the need to cover the area uniformly. The best uniformity is an average-to-minimum ratio of 3:1 or less. For less critical areas this can slip to 6:1. The maximum- to minimum-illumination ratio should be kept to less than 10:1 if visible nonuniformity is to be avoided.

Bibliography

American National Standard Practice for Roadway Lighting, ANSI/IES RP-8, Illuminating Engineering Society of North America, New York, 1977.

"IES Approved Method for Photometric Testing of Roadway Luminaires," *Journal of the Illuminating Engineering Society,* January 1978, p. 132.

IES Lighting Handbook, 5th ed., sec. 20, Roadway Lighting Illuminating Engineering Society, New York, 1972.

Lighting Photometric Data and Sample Calculations, GET-3100G, General Electric Company, Lighting Systems Business Department, Hendersonville, N.C., 1976.

12

Design Procedures for Roadways, Parking Lots, and Outdoor Areas

Design procedures for lighting outdoor areas with roadway, area, and high-mast luminaires are somewhat different from those used in floodlighting. The designer can apply photometric data directly to the area, without making allowances for changes in vertical aiming angles, which are necessary when floodlights are used.

The preceding chapter gave background information needed before a lighting system using roadway, area, or high-mast luminaires can be designed. This chapter will discuss actual design problems. Included are design procedures for roadways and parking lots and a section on high-mast lighting of outdoor areas.

Illumination Levels

Average maintained illumination levels of roadways are based on the light output of the luminaire at the time of relamping. This is the lowest illumination level that the system will produce before it is cleaned and relamped. The Illuminating Engineering Society of North America (IES) does not give specific recommendations for roadways which are within an industrial complex. However, the recommendations given in Table 12-1 for average maintained illumination levels are based on the guidelines which govern the selection of light levels for city streets and highways.

Intersections or converging points require levels of illumination higher than those given in the table. These can be attained by putting the first luminaire in each direction at the intersection. The intersection will thus receive contributions from both streets and will have a level of illumination close to the sum of the levels in both streets.

Sample Roadway Problem

Assume that a lighting system must be designed for a roadway that will be heavily traveled during shift changes. Reading from Table 12-1, the required average illumination level is 21.52 lux (2 footcandles). The proposed mounting height is 9.14 meters (30 feet), and the street width is

Table 12-1 Recommended levels of illumination for roadways

Major arteries where heavy traffic will exist during the night or during shift changes 21.5 lux (2 footcandles)—uniformity 3:1
Collector roads connecting parts of an industrial area to the main roads 12.9 lux (1.2 footcandles)—uniformity 3:1
Alleys or roads with light traffic and pedestrian walkways 6.5 lux (0.6 footcandles)—uniformity 6:1

9.14 meters. The mounting height is high in relation to the street width, because the street is so narrow. The height is needed in order to project light further down the street than would be possible with shorter poles. Assume that the luminaires are mounted over the curb.

A 250-watt high-pressure sodium (HPS) source mounted in a medium, semicutoff, type III luminaire has been tentatively chosen for this application. This is the same luminaire described in the previous chapter. A low-wattage source is needed because of the low mounting height, and HPS luminaires have been chosen because of their high efficiency.

The selection of the best lamp size and photometric distribution can be done only by trial and error. A lamp size should be chosen that does not require a pole spacing much over 5 times the luminaire mounting height. Too wide a spacing will make it impossible to obtain 3:1 average-to-minimum uniformity. The designer can select the best luminaire type by comparing the different photometric distributions in order to determine the best compromise between utilization and uniformity.

SPACING

Spacing can be determined from the following formula:

$$S = \frac{\text{LL} \times \text{CU} \times \text{LLF}}{E \times W}$$

where LL = initial lamp lumens
CU = coefficient of utilization
LLF = light-loss factor
E = average maintained illumination level
and W = street width

The initial lumen rating for the lamp selected is 27,500 lumens. The coefficent of utilization (CU) can be read from the utilization curve given in Fig. 12-1. The ratio of transverse distance to mounting height is 9.14/9.14 = 1.0. Find the 1.0 point on the lower coordinate, and go up to the street-side curve. Read the utilization, which in this case is 42 percent.

The light-loss factor compensates for the loss of illumination that occurs as an installation becomes older. It is made up of a number of components, the most important of which are luminaire dirt depreciation (LDD) and lamp lumen depreciation (LLD). The accumulation of dirt on luminaires results in a loss of light output and therefore a loss of light on the roadway. The LDD is a measure of that loss. It is a function of the kind of environment, volume of traffic, mounting height, degree of luminaire gasketing, and time since last cleaning. Sealed and filtered luminaires do not depreciate as fast as those which are not filtered.

The LLD indicates the light output of an average lamp in relation to its operating time. It can be expressed as the lumen output at the end-of-relamping period, or as the mean value, which is sometime around the midpoint of the lamp's life. In roadway lighting, the LLD is the percentage of the initial lamp lumens that the lamp will be producing when it is replaced; this is near to the end of rated life, and so that end-of-relamping value is used. Since streetlighting equipment must be serviced from a bucket truck or other type of mobile platform, it is usually more economical to group-relamp the system after a specific burning period than to randomly replace lamps as they fail.

Figures for LDD and LLD are given by manufacturers. They have been verified by actual tests. For the luminaire used in the sample problem, the LLD is 0.73 and the LDD is 0.95, since it is a sealed and filtered luminaire. Light-loss factor is LLD × LDD and is therefore 0.73 × .095 = 0.69.

The illumination level for the example is 21.5 lux (2 footcandles) and the street width 9.14 meters (30 feet). Substituting in the formula,

$$S = \frac{\text{LL} \times \text{CU} \times \text{LLF}}{E \times W}$$

$$S = \frac{27{,}500 \times 0.42 \times 0.69}{21.5 \times 9.14} = \frac{7969.5}{196.5} = 40.56 \text{ meters (133 feet)}$$

The results are the same if footcandles and feet are used for E and W: 27,500 × .042 × 0.69/2 × 30 = 7969.5/60 = 132.8 feet (40.5 meters).

Since pole height is 9.14 meters, the ratio of longitudinal spacing to mounting height (S/MH) is 40.56/9.14 = 4.4 (or if the values are expressed in feet, 132.8/30 = 4.4).

UNIFORMITY

The total lumens on the street is not the only consideration, though, as explained in the previous chapter. Lighting uniformity is as important as the illumination level. Uniformity is the ratio of the average level of illumination (21.5 lux in this case) to the minimum level. It should be kept as close to 3:1 as possible.

The designer can determine actual uniformity from the illumination data for the luminaire to be used. This information is given as part of the photometric data. Figure 12-1 includes the illumination data for the 250-watt HPS luminaire that is used for the sample problem. More isoillumination lines and the two curb lines of the roadway in the example have been noted directly on the data. One curb line is directly in line with the luminaire position. The other is at 1.0 MH, which is the ratio of the 9.14-

meters (30 feet) road width, or transverse distance, to the 9.14-meter mounting height.

There will be contributions on the roadway from more than one luminaire. Figure 12-2 includes a layout of the roadway with three contributing luminaires, spaced at a distance of 40.5 meters, or 4.4 MH. There is no way to determine in advance where minimum levels will occur, but points that ought to be checked have been noted on the drawing. They lie

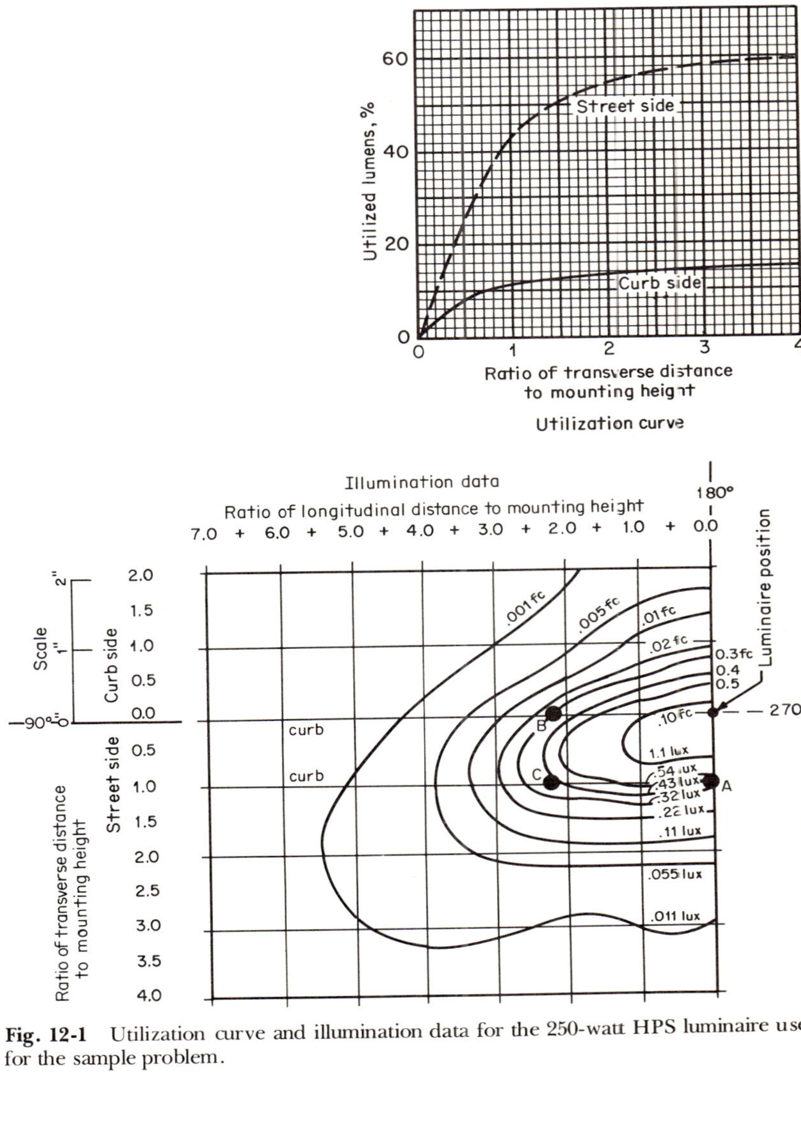

Fig. 12-1 Utilization curve and illumination data for the 250-watt HPS luminaire used for the sample problem.

	Point A		Point B		Point C	
	TD/MH	LD/MH	TD/MH	LD/MH	TD/MH	LD/MH
Luminaire number 1	9.1/9.1 = 1	40.5/9.1 = 4.4	0/9.1 = 0	20.2/9.1 = 2.2	9.1/9.1 = 1	20.2/9.1 = 2.2
Illumination level, lux (footcandles)	0.033	(0.003)	0.32	(0.03)	0.43	(0.04)
Luminaire number 2	9.1/9.1 = 1	0/9.1 = 0	0/9.1 = 0	20.2/9.1 = 2.2	9.1/9.1 = 1	20.2/9.1 = 2.2
Illumination level, lux (footcandles)	0.43	(0.04)	0.32	(0.03)	0.43	(0.04)
Luminaire number 3	9.1/9.1 = 1	40.5/9.1 = 4.4	0/9.1 = 0	60.7/9.1 = 6.6	9.1/9.1 = 1	60.7/9.1 = 6.6
Illumination level, lux (footcandles)	0.033	(0.003)	—		—	
Total illumination level (E), lux (footcandles)	0.496	(0.045)	0.64	(0.06)	0.86	(0.08)
$E \times 27{,}500$ (lamp lumens)/1000 lux (footcandles)	13.64	(1.2)	17.6	(1.6)	23.65	(2.1)
$E \times 27.5 \times 0.69$ (light loss factor), lux (footcandles)	9.4	(0.85)	12.1	(1.1)	16.3	(1.5)

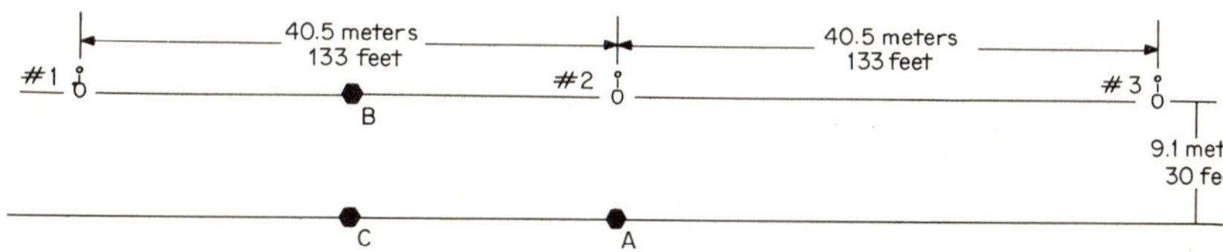

Fig. 12-2 Minimum light levels may occur at points A, B, or C. These are good points for the designer to check when determining the uniformity of a system, as has been done here for the sample problem.

directly across from a luminaire (A) and midway between luminaires on each side of the street (B and C).

Before checking these points by using the illumination data, the designer must express dimensions in both the transverse and longitudinal directions as multiples of the mounting height. To ensure accuracy, it is easiest to do this in chart form, as in Fig. 12-2. Illumination data for points A, B, and C for each of the three luminaires have also been transferred to the chart. In each case, the figures have been read from illumination data in Fig. 12-1 by finding the point where the transverse distance/mounting height (TD/MH) and the longitudinal distance/mounting height (LD/MH) meet. Luminaire position is the placement of luminaire number 2.

Illumination levels from each of the three luminaires have been added together to find the total level at each of the three points. Since these levels are given per 1000 lumens, they must be converted by multiplying each one by the total lumen output of the luminaire (27,500) and dividing by 1000. Thus, the illumination level at point A is 0.496 × 27,500/1000 = 0.496 × 27.5 = 13.64 lux.

This figure must then be multiplied by the light-loss factor to yield the maintained level of illumination on the roadway at the associated point: 13.64 × 0.69 = 9.4 lux at point A, for instance.

The minimum can occur at any point; its location will vary with the width of the street, spacing, mounting height, overhang, and photometric distribution of the luminaire. In this case it occurred directly across the street, since point A has a light level of 9.4 lux.

The uniformity ratio for this roadway is found by dividing the average light level (21.5 lux) by the minimum level (9.4 lux) 21.5/9.4 = 2.3. The uniformity ratio is 2.3:1, which is so close to the 3:1 requirement that it is doubtful whether the system could be substantially improved. Raising the pole height would improve the uniformity but would also reduce utilization and add to costs.

If the uniformity is better than 2:1, a higher-wattage lamp or shorter pole should be investigated. A higher-wattage source would allow the designer to increase the spacing, which would lower system cost. A shorter pole would also reduce system cost.

Since uniformity ratios for the sample problem are within reasonable limits with one-side spacing, there would seem to be no reason in this case to use staggered spacing, which improves uniformity but adds to wiring costs.

Area Lighting with Roadway, Area, and High-Mast Luminaires

Spacing-to-mounting-height ratios of approximately 4.5:1 are recommended for area lighting with roadway, area, or high-mast luminaires, because of their photometric distribution. Most are designed to deliver maximum candlepower at a vertical angle of 70 to 75 degrees. Above the point of maximum candlepower, light output decreases rapidly. With cutoff luminaires, there is very little light or no light at all above 90 degrees, and there is very little light between 80 and 90 degrees with most standard roadway, area, and high-mast systems.

Figure 12-3 shows a typical outdoor area lighted with roadway luminaires placed on four poles in a square pattern. Spacing between poles is 4.5 MH. Point A is located in the center of the square. This is the most difficult spot in this area to light; maximum candlepower from the luminaire should therefore be directed at a vertical angle of 72.5 to 74.2 degrees. The greatest candlepower will then reach point A.

Chapter 9 describes the candlepower requirements for lighting a point at a distance from the pole as compared with lighting a point directly beneath the pole. It takes 33 to 48 times as much candlepower to light point A to a given level of illumination as it does to provide the same level directly beneath the pole.

Point B is midway between two poles and can be reached at an angle of 66 to 68 degrees. It takes 16 to 19 times as much candlepower to light this spot to a given level as it does to provide the same level directly beneath the pole.

In order to expand S/MH ratios above 4.5:1, maximum luminaire candlepower has to be above 75 degrees vertical. The resulting luminaire brightness produces a very glaring installation. If widely spaced shielded or louvered luminaires are used to reduce brightness, there will be dark areas between poles and uniformity will be poor.

A horizontal spacing of 4.5 MH allows for good uniformity without

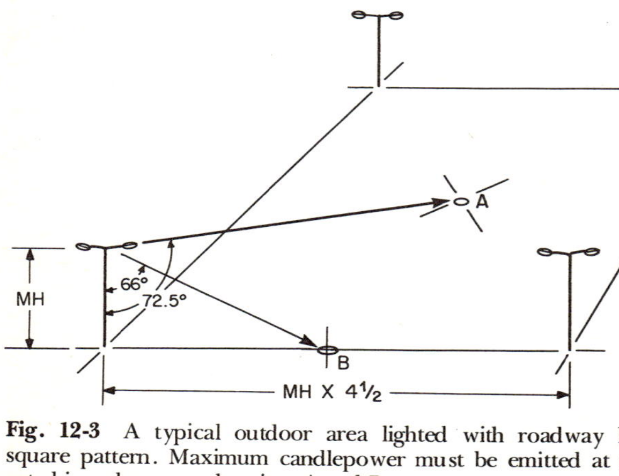

Fig. 12-3 A typical outdoor area lighted with roadway luminaires in a square pattern. Maximum candlepower must be emitted at the angles indicated in order to reach points A and B.

excessive luminaire brightness. Wider spacing is possible, but only if the minimum light level is unimportant.

Parking Lots

Parking lots are lighted in the same manner as other outdoor areas, except that poles must be placed out of the path of moving traffic. Normally, parking lots are divided into 18.3-meter (60-foot) units with two 6.1-meter (20-foot) parking lanes plus a 6.1-meter access lane. Entrances, exits, and walkways can alter this pattern somewhat.

The line of poles in one direction, then, will be placed on centers in multiples of 18.3, 36.6, 54.9, 73.2 meters, etc.—depending on pole height and the size of the lot.

If there is more than one row of poles, spacing in the other direction will depend on the width of the lot. Each pole can effectively light 2 to 2.25 MH in any direction, since spacing between poles can be 4.5 MH. This means that poles need not be placed any closer to the edge of a lot than 2 MH. It is desirable to space the poles in a square pattern, but the dimensions of the parking lot will dictate the actual spacing.

If the parking lot is irregularly shaped, it can be divided into smaller areas for design purposes. Roadway, area, or high-mast luminaires can be used, as can combinations of these systems. Floodlights are sometimes effective where pole locations are limited. Their use is covered in Chaps. 9 and 10. In general, floodlights will have poorer utilization and result in increased glare.

For a sample problem, assume that a parking area 122 by 244 meters (400 by 800 feet) must be lighted. This can be done with either high-mast poles or rows of shorter poles in a grid pattern.

LIGHTING WITH ONE ROW OF POLES

1. Select the light level. The IES recommends 10.76 lux (1 footcandle) for self-service parking lots. This also applies to industrial lots, although a higher level would be recommended if television surveillance were to be used. (See Chap. 13.)

2. Choose a tentative pole height, based on the width of the lot, which in this case is 122 meters (400 feet). The width that can be lighted with one pole placed in the center is 4 times the mounting height (since each pole can light 2MH in any direction), and so 30.5-meter (100-foot) poles could be used. Any local restrictions on pole height should be checked. Since it would be difficult to service the luminaires at 30.5 meters, a high-mast lowering device should be considered.

3. Measure the length of the parking area, which in this case is 244 meters (800 feet), and note the location of the marked parking areas. Figure 12-4 gives the layout of the lot used for the example. Poles must be between parking lanes, and no pole should be further than 2 MH from the perimeter of the lot.

4. Using the layout, estimate the number of poles needed to cover both length and width, and their placement. Poles 30.5 meters tall can be placed 4.5 MH , or 137.2 meters (450 feet), apart.

If this were an open area, 137.2 meters would be good spacing. Since it is a parking lot, however, spacing is affected by the layout of the lot. The closest figure to 137.2 meters (450 feet) which is a multiple of 18.3 meters (60 feet) is 128 meters (420 feet). However, because of the way this lot is designed, 134-meter (440-foot) spacing is a better choice, as shown in Fig. 12-4. This spacing leaves 55 meters (180 feet) on either end of the lot, which is less than 2 MH ($2 \times 30.5 = 61$ meters) and therefore acceptable.

The poles can be placed in the other direction 61 meters (200 feet) from each side of the 122-meter (400-foot) lot, since each pole can effectively

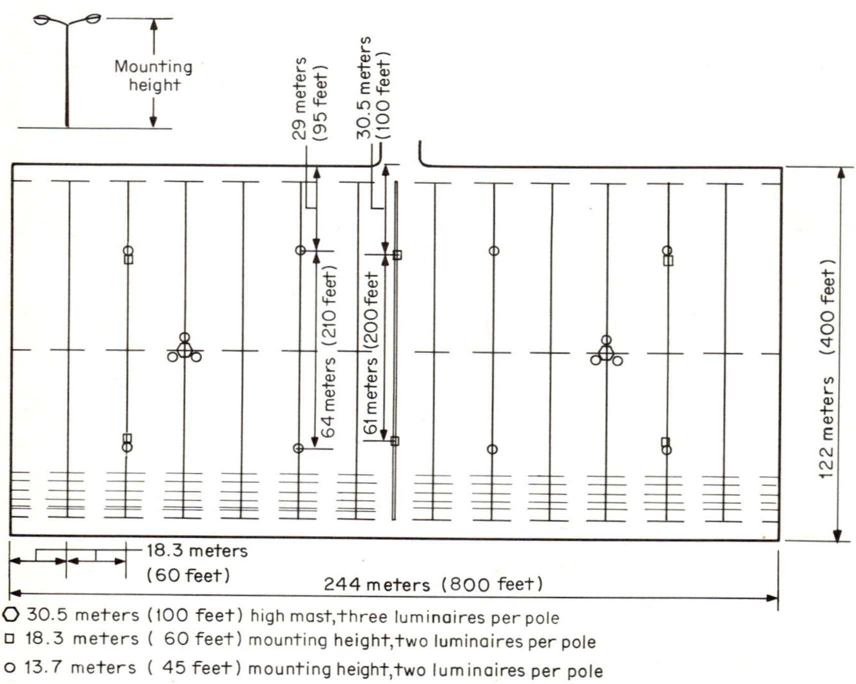

Fig. 12-4 Layout of the parking lot used in the sample problem and lighted with three different systems.

distribute light a distance of 2 to 2.25 MH. Two 30.5-meter (100-foot) poles can thus be used to light this lot (see Fig. 12-4).

5. Make a tentative luminaire choice. The most economical system will be the one that uses the fewest luminaires and has the lowest power requirements but still meets the required level of illumination. A 1000-watt HPS lamp in a high-mast luminaire is the most efficient system, and so this should be tried first.

6. Use the formula

$$E = \frac{N \times \text{LL} \times \text{CU} \times \text{LLF}}{\text{area}}$$

to determine the total number of luminaires (N) needed to light the lot to a maintained level (E) of 10.76 lux (1 footcandle). The lumen rating (LL) for the 1000-watt HPS luminaire is 140,000.

The CU for the luminaire can be read from Fig. 12-5. For an area 2 MH wide, the CU is 29 percent on the street side. On the house or curb side it is also 29 percent, since the light pattern of the luminaire is type V and symmetrical. The total utilization is therefore 29 + 29 = 58 percent.

The utilization curve is based on lighting a long horizontal area. In this example however, the area ends 2 MH from the pole base in one direction; some light which is included in the utilization value will spill over one end of the lot. If the pole is located 2 MH from the edge, though, the error in the utilization figure will be less than 10 percent which is within normal design accuracy.

The LLD in this case is 0.9, and the LDD is 0.95, since the luminaire is sealed and filtered. The light-loss factor (LLF) is therefore 0.9 × 0.95 = 0.855. The area of the parking lot is 122 × 244 = 29,768 square meters (400 × 800 = 320,000 square feet). Solving for N,

$$10.76 = \frac{N \times 140{,}000 \times 0.58 \times 0.855}{29{,}768}$$

$$10.76 = \frac{N \times 69{,}426}{29{,}768}$$

$$69{,}426\,N = 320{,}304$$

$$N = 4.6$$

Since two poles will be used, this will result in 4.6/2 = 2.3 high-mast luminaires per pole. The final design will therefore use three luminaires per pole. This will produce approximately 14 lux (1.3 footcandles), but it is the closest luminaire quantity that still meets the required illumination level.

7. Check for possible minimum-illumination points, as indicated in Fig. 12-6. Figure 12-5 reproduces the isoillumination curves for the 1000-watt

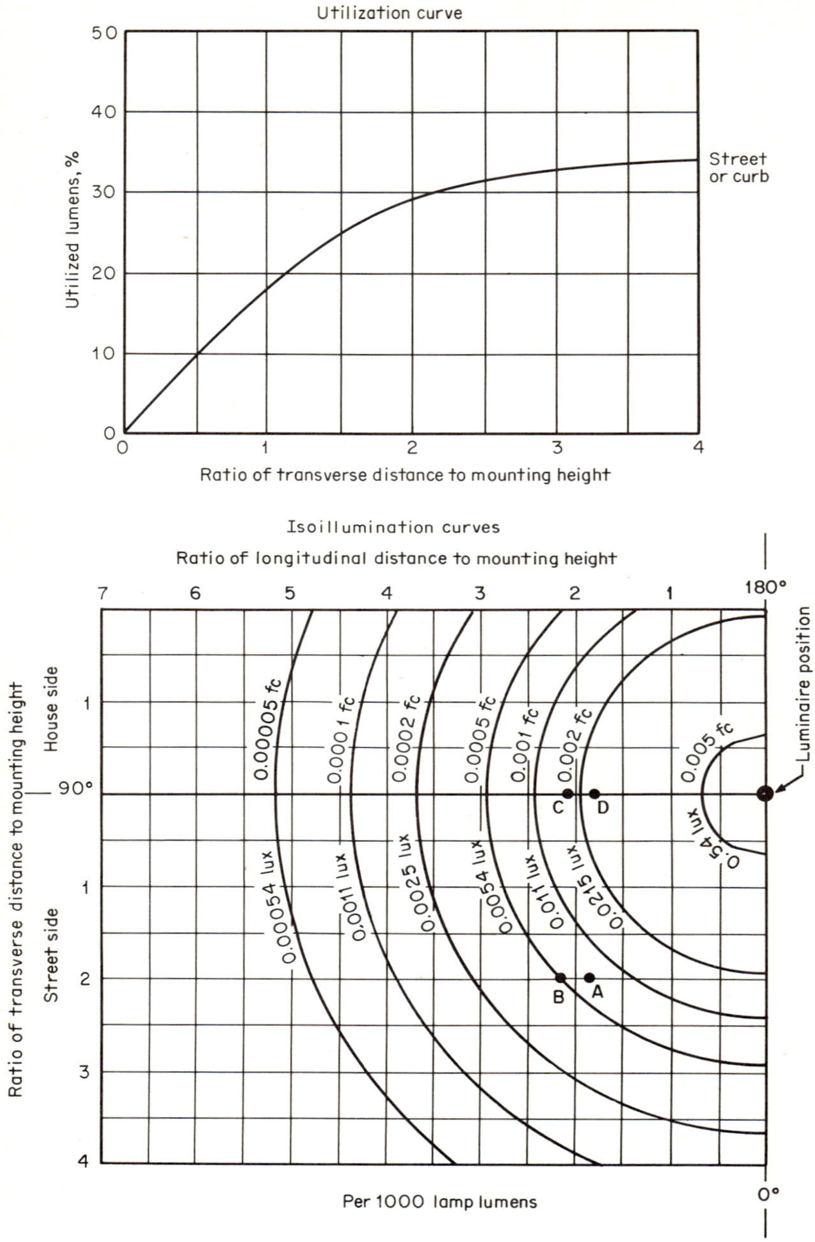

Fig. 12-5 Utilization curve and isoillumination curves for the 1000-watt HPS luminaire used in the sample problem. Levels of illumination at points A, B, C, and D in Fig. 12-6 can be read from the curves.

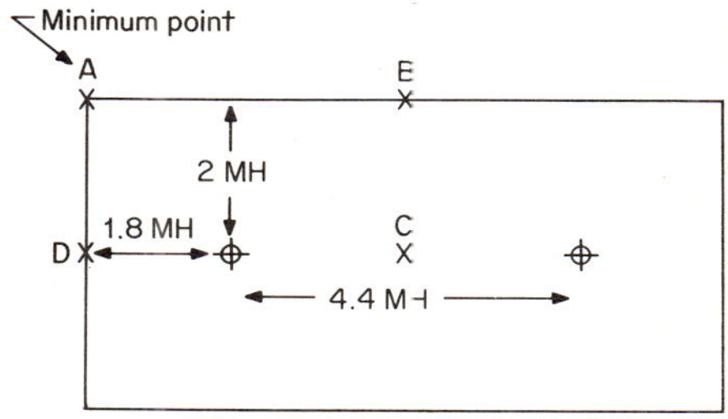

Fig. 12-6 Points in the parking lot which might have minimum levels of illumination.

HPS luminaire that is being used. Points A, B, C, and D from Fig. 12-6 are indicated on the diagram. The data is based on a 30.5-meter mounting height, which is the mounting height in the sample problem. Therefore, a correction factor is not needed.

Only one pole holding three luminaires will be contributing to the light level at point A, in the extreme corner of the lot. Figure 12-5 indicates that about 0.0075 lux (0.0007 footcandle) will be directed to this spot by each luminaire. This level is expressed per 1000 lamp lumens or per 140,000/1000 = 140. The maintained light level is thus 0.0075 × 140 × 3 (number of luminaires) × 0.855(LLF) = 2.69 lux (0.24 footcandle).

Point B will receive about 0.0054 lux from each luminaire. Light from the two poles will contribute to point B; the maintained light level is therefore 2 ×0.0054 × 140 × 3 × 0.855 = 3.88 lux, or 0.35 footcandle.

The light level at point C will be 2 ×0.0172 × 140 × 3 × 0.855 = 12.35 lux (1.1 footcandles).

The lowest level of illumination will thus be at point A. The level at that point, 2.69 lux (0.24 footcandle), is about 1/5 the average level of 13.99 lux (1.3 footcandles); the uniformity is therefore 5:1. This occurs only in the extreme corners and is not enough reason to change the lighting system, since the bulk of the lot will have better than 3:1 uniformity.

LIGHTING WITH MORE THAN ONE ROW OF POLES

The 122- by 244-meter parking lot used in the previous example can also be lighted with more than one row of shorter poles. The following pro-

Correction factors for mounting heights other than 30.5 meters (100 feet)

MOUNTING HEIGHT							
Meters	24.4	27.4	30.5	33.5	36.6	39.6	42.7
Feet	80	90	100	110	120	130	140
MULTIPLIER	1.56	1.23	1.0	0.83	0.69	0.59	0.51

cedure can be used to determine the number of luminaires and poles needed:

1. Pick the illumination level, in this case a maintained level of 10.76 lux (1 footcandle).

2. Choose the tentative quantity, location, and height of the poles. From the previous example it can be seen that the lot can be effectively lighted with two 30.5-meter (100-foot) poles. If two rows of poles are used, the distance to the edge should still be approximately 2 MH. The distance between poles can be 4.5 MH. The lot width can thus be divided into segments of 2, 4.5, and 2 MH. Divide the width in meters by the MH figure to find pole height: 122/(2 + 4.5 + 2) = 122/8.5 = 14.35 meters (47 feet).

Next investigate the 244-meter (800-foot) length of the lot. Given a tentative pole height of 14 meters, it would take four poles to cover the length: 244/(2 + 4.5 + 4.5 + 4.5 +2) = 244/17.5 = 13.9 meters (46 feet). (Refer to Fig. 12-4 for clarification of this procedure.)

A 13.7-meter (45-foot) pole would be good tentative choice for this lot. If desired, pole height could be increased with a 0.6-meter (2-foot) concrete base. This is an inexpensive way of increasing pole height and also eliminates the need for added protective guards for poles mounted inside a parking lot. For the example, however, assume a 13.7-meter pole height.

For comparison purpose, calculate the pole height needed for two rows of three poles each. If three poles were used, the height would have to be 244/(2 + 4.5 + 4.5 + 2) = 244/13 = 18.8 meters (62 feet).

An 18.3-meter (60-foot) pole would be sufficient, and so this lot can be laid out with either two rows of three 18.3-meter poles or two rows of four poles at 13.7 meters.

3. Position the poles on the layout of the parking area (Fig. 12-4). In the case of either the 13.7-meter pole or the 18.3-meter pole, the first row lengthwise will have to be 36.6 meters (120 feet) from the edge of the lot, because of the parking areas. This is over 2 MH for the 13.7-meter pole, but will probably produce acceptable results if care is taken in positioning the luminaires.

The spacing between poles will be 54.9 meters for the 13.7-meter pole, or 4 MH. Spacing will be slightly higher between the second and third poles because of the center access road.

For the 18.3-meter pole, the spacing is half the lot length of 122 meters minus the distance from the edge of the lot, or 122 − 36.6 = 85.4 meters. This is a spacing of 4.7:1, and is acceptable.

Across the 122-meter (400-foot) dimension, each pole has been placed 29 meters (95 feet) from the edge, which is slightly over 2 MH for the

13.7-meter pole and under 2 MH for the 18.3-meter pole. The distance between poles in either case is 64 meters. This is a spacing of 4.7:1 for the 13.7-meter pole and 3.5:1 for the 18.3-meter pole.

Moving in the 18.3-meter poles 1.5 meters (5 feet) from either edge would improve the luminaire utilization without sacrificing uniformity. Actual utilization figures can be read from photometric data.

4. Determine a tentative luminaire quantity and lamp size for each suggested pole height. In general, type III or type IV roadway or area luminaires are used for area lighting. Earlier in this chapter, Fig. 12-1 gave the utilization curve for a typical type III roadway luminaire using a 250-, 310-, or 400-watt HPS lamp. At least two luminaires per pole, mounted back to back, are required with type III distribution in order to get a symmetrical pattern of light. In this example, first try 2 luminaires per pole, which would be 16 luminaires for the 8-pole, 13.7-meter (45-foot) system, and 12 for the 6-pole, 18.3-meter (60-foot) system.

Using these as tentative number of luminaires, solve for lamp lumens with the following formula:

$$E = \frac{N \times \text{LL} \times \text{CU} \times \text{LLF}}{\text{area}}$$

The illumination level is again 10.76 lux (1 footcandle). The CU can be read from Fig. 12-1. It is based on the use of the luminaire on a long roadway. In this case, since all the luminaires are mounted within 2 MH of the edge, some of the light will be lost. The utilization will lie somewhere between 75 percent (60 percent street-side, 15 percent curb-side) and 70 percent (55 percent street-side, 15 percent curb-side). It is not necessary to be absolutely precise at this point, and so an approximate CU of 72 percent can be used.

Because of their high efficacy, HPS lamps will be used. All lamp sizes have the same LLD of 0.9, and a sealed and filtered luminaire has an LDD of 0.95. The light-loss factor is thus LLD ×LDD = 0.9 × 0.95 = 0.855.

The area is 122 × 244 = 29,768 square meters.

Assuming 16 luminaires for the system using 13.7-meter (45-foot) poles, substitute the formula:

$$10.76 = \frac{16 \times \text{LL} \times 0.72 \times 0.855}{29{,}768}$$

$$320{,}304 = 9.8 \text{ LL}$$

$$\text{LL} = 32{,}684$$

Table 2-1 in Chap. 2 gives a chart of wattages and lamp lumens. The 310-watt HPS lamp having 37,000 lumens would be a good choice for this application.

The system using 18.3-meter (60-foot) poles would use 12 luminaires.

Substituting in the formula,

$$10.76 = \frac{12 \times \text{LL} \times 0.72 \times 0.855}{29{,}768}$$

$$320{,}304 = 7.4 \text{ LL}$$

$$\text{LL} = 43{,}284$$

Table 2-1 shows that a 400-watt HPS lamp having 50,000 initial lamp lumens would be required.

5. Orient the luminaires. A roadway luminaire has a wider distribution perpendicular to the orientation of the luminaire. Figure 12-7 gives the photometric curves for the HPS luminaire used in this example. Since two luminaires are being used back to back, the curves for both luminaires have been joined. As can be seen, the light pattern produces the same illumination level at a distance of about 3 MH and angles of 90 and 270 degrees as it does at a distance slightly less than 2 MH and an angle of 0 degrees.

This is typical for roadway luminaires of types I through IV. Only a type V would produce the same light level at a given distance in front of the luminaire (0 degrees) as it does at the same distance to either side (90 and 270 degrees).

When the luminaires are installed, they should be oriented so that the maximum coverage coincides with the greater dimension that the luminaire needs to cover. This may or may not coincide with the greater dimension in the general area being lighted.

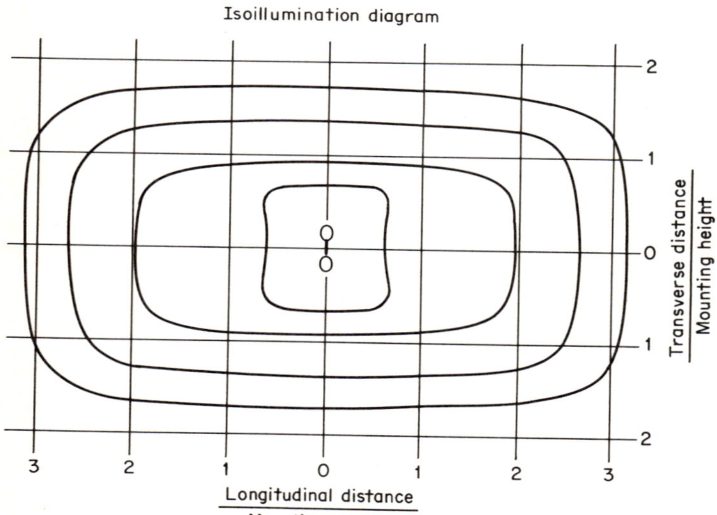

Fig. 12-7 Photometric curves for two luminaires have been joined together to show the actual light pattern in the parking lot used in the sample problem.

6. Check for minimum light levels and uniformity ratios as explained previously for the high-mast system.

7. Choose the system that will be used. There is no simple way of doing this. The costs of the poles and foundations, the initial system cost, and the projected operating and maintenance costs are all factors. Light levels, uniformity, and utilization also should be considered. In this example, since both systems provide acceptable lighting, the 18.3-meter pole would probably be the better choice, since it would require fewer poles and fewer luminaires.

ALTERNATE SYSTEM USING AREA TYPE V LUMINAIRES

There are a number of parking-lot luminaires that have a circular or type V distribution. This type of luminaire is designed for commercial parking lots where its daytime appearance is an important consideration. Its efficiency and beam characteristics are similar to those of high-mast luminaires. These luminaires are also available in the same range of lamp sizes, from 250 to 1000 watts. In the previous example, six 1000-watt luminaires could be used on 18.3-meter poles to produce the same light level as was produced by the other systems. These luminaires tend to be more expensive than roadway luminaires, but because of the type V distribution, only one luminaire per pole is required to obtain the necessary coverage.

Graphical Solution to High-Mast Lighting Problems

Figure 12-8 is a graphical method of determining combinations of pole height, luminaire type, number of luminaires, and initial lux or footcandle levels. It is used in combination with photometric data for the proposed luminaire choice.

Also in Fig. 12-8 is a drawing showing the relation between the vertical angle of the candlepower from a high-mast luminaire and the distance this light will be projected from the pole. Distances are given as ratios of horizontal distance to mounting height.

The large graph is a means of solving the equation

$$E = \frac{\text{cd} \cos^3 \theta}{\text{MH}^2}$$

where E = level of illumination at a point, in lux or footcandles
cd = candela value of luminaire at any vertical angle θ
θ = vertical angle at which light is directed
MH = luminaire mounting height, which can be assumed to be the pole height (unless the pole is mounted on a raised base)

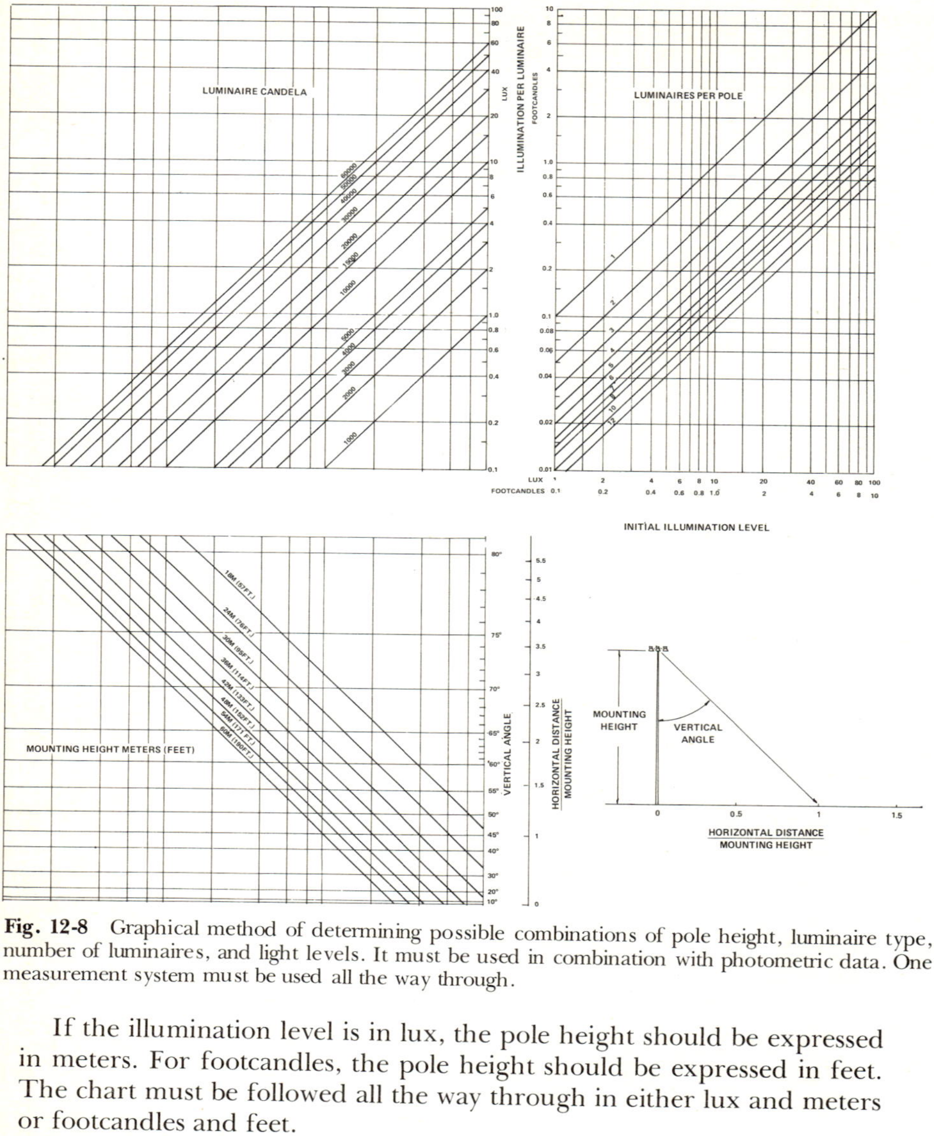

Fig. 12-8 Graphical method of determining possible combinations of pole height, luminaire type, number of luminaires, and light levels. It must be used in combination with photometric data. One measurement system must be used all the way through.

If the illumination level is in lux, the pole height should be expressed in meters. For footcandles, the pole height should be expressed in feet. The chart must be followed all the way through in either lux and meters or footcandles and feet.

The purpose of the chart is to find a starting point for designing a high-mast system. Results are estimates and must be verified. This can be done by using the formula given above or by consulting manufacturers who offer computer services.

The horizontal distance from the pole base to the point being measured, as well as a tentative pole height, should be determined before the graph is used. Then, when the ratio of the horizontal distance to the mounting height is known, the vertical angle at which the light is directed can be read on the right side of the lower left section of Fig. 12-8. For instance, if the point being measured is 107 meters from a 31-meter pole, the ratio is 3.5:1 and the vertical angle approximately 74 degrees. The right side of the lower chart is a plot of the cosine cubed of an angle, on a logarithmic scale.

The diagonal lines in the lower section show the mounting height of the luminaire in meters first and then in feet. This chart is plotted so that the numbers on the top represent $\cos^3 \theta/MH^2$.

The upper left section of Fig. 12-8 is a plot of luminous intensity in candelas. The right scale is the solution of the equation and shows the lux level per luminaire for a particular situation. The left ordinate of the right graph gives illumination per luminaire, in footcandles. Numbers opposite each other are not equivalent.

The diagonal lines in the upper right section of the chart give the number of luminaires per pole. The scale on the bottom of this section gives the total initial illumination level at a point for a particular system, in lux or footcandles.

Depending on what is known or what is needed, the designer can start anywhere on the chart and work in any direction.

DESIGNING FOR MINIMUM ILLUMINATION LEVELS

Most high-mast lighting systems are designed to deliver light everywhere in the area. The design of the system must satisfy two values: the average illumination level and the minimum illumination level. It is usually easier to design for the minimum level and let the average float. In general, if the mimimum level is satisfied, the average level will be at least 3 times higher, because of the photometric characteristics of high-mast luminaires. This provides 3:1 average-to-minimum uniformity, which is an acceptable level.

The corner of an area is usually a point of low illumination when a high-mast system is used. Figure 12-8 can be used to design the system so that the minimum light level will be provided at that point.

SAMPLE PROBLEM

For a sample problem, assume that a single pole is to be mounted in the center of a 152.4- by 152.4-meter (500-by 500-foot) storage area. A television surveillance system will be used, and the minimum maintained light level that the camera will tolerate is 5.38 lux (0.5 footcandle). Using meters and lux, determine what pole height and how many luminaires to use.

Step 1. Make a tentative luminaire choice. Since 1000-watt HPS sources have the highest efficacy, a luminaire using this source should be considered first. For this example, the same 1000-watt luminaire used in the previous examples will be used.

Step 2. Convert the minimum maintained illumination level of 5.38 lux to an initial value by using the light-loss factor. The LLD for this lamp is 0.9 and the LDD 0.95, since a sealed and filtered luminaire will be used. Therefore the initial value should be $5.38/(0.9 \times 0.95) = 6.29$ lux.

Step 3. From manufacturer's data, find the maximum number of luminaires that can be mounted on one pole. It is usually 12, and so 12 will be considered the maximum for this example. It is a good idea to try first to use the maximum number of luminaires per pole, since this saves on pole costs.

Step 4. Determine the maximum candela point for this luminaire and the limit of its range. This can be done by referring to the photometric data in Table 12-2. The maximum candela value is 212 candelas per 1000 lumens; this occurs at 60 degrees. The candela value drops to 102 candelas per 1000 lamp lumens at 75 degrees. This, then, is the upper limit of the useful beam, since the candela value is less than half the maximum.

Step 5. Convert the candela values per 1000 lumens to actual lumens for angles from the maximum candela point to the limit of the useful beam, or in this case from 60 to 75 degrees. Since the 1000-watt HPS

Table 12-2 Photometric data for a 1000-watt HPS luminaire

Max. Cone 60
Max. Candela 212

Cone	*Candela*
0°	70
5°	69
15°	61
25°	69
35°	90
45°	129
55°	191
60°	212
65°	186
70°	148
75°	102
80°	63
85°	26
90°	10

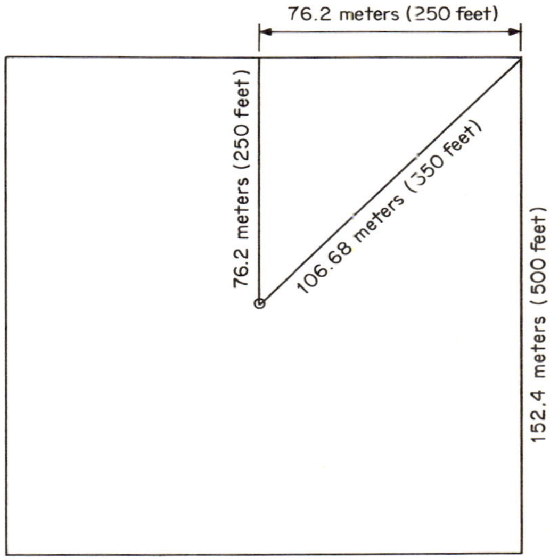

Fig. 12-9 Distances to the perimeter and the corner of the sample area.

source has 140,000 lamp lumens, the values per 1000 lumens must be multiplied by 140 to yield actual values.

60 degrees: 211 × 140 = 29,680 candelas
65 degrees: 186 × 140 = 26,040 candelas
70 degrees: 148 × 140 = 20,720 candelas
75 degrees: 102 × 140 = 14,280 candelas

Step 6. Determine the pole height required for the limit of useful candlepower, or candlepower at 75 degrees, to reach the corner of the area, a possible point of low illumination. This would be an ideal situation. Figure. 12-8 shows that 75 degrees occurs at 3.7 MH. A pole height can be determined, then, since the distance from the pole base to the corner of the lot is known; Fig. 12-9 shows that this distance is 106.7 meters. If this distance is set equal to 3.7 MH, the pole height can be found by dividing: 106.7/3.7 = 28.8 meters. A 30-meter pole can therefore project the edge of the useful beam to the corner of the area.

Step 7. Refer to the large chart in Fig. 12-8 to see if this pole height can be used. Use the following procedure:

1. On the right part of the lower left portion of the graph, find the 3.7-MH and 75-degree point.

2. Draw a line to the left from there to the 30-meter pole height.

3. From this point, draw a line straight up to the candela value at 75 degrees. Since it is 14,280, use the 15,000-candela line.

4. From there, draw a line to the right to the slanted line indicating 12 luminaires per pole (the maximum).

5. Draw a line down from that point to read the initial lux level for this situation. In this case, 12 luminaires will deliver slightly less than 4 lux, which is below the required minimum level. This, then, is not an acceptable solution.

Step 8. Repeat the process, assuming a 70-degree vertical angle will reach the corner of the lot. Seventy degrees is 2.7 MH so 106.7/2.7 = 39.5 meters. Since there is no line on the graph for a 39.5-meter pole, draw a line from 70 degrees to the appropriate spot between lines. From this point, go up to 20,720 candelas (the candelas at 70 degrees) and over to 12 luminaires. This system will deliver better than 7 lux initially to the corner of the lot. The required initial lux level is 6.29, and so a 39.5-meter pole would probably be a satisfactory choice.

Step 9. For comparison purposes, determine the solution that uses the fewest luminaires. This occurs when the maximum luminaire candelas coincides with the minimum illumination point—in this case, the corner of the lot, 106.7 meters away. The maximum candelas occur at 60 degrees, which is 1.7 MH. The possible pole height is then 106.7/1.7 = 62.8 meters: a 60-meter pole could be used. Repeating the procedure throughout the graph shows that six luminaires on a 60-meter pole would deliver 6.3 lux to the corner of the lot, which is the required level.

Step 10. Determine the average illumination level and uniformity for both the 39.5-meter pole and the 60-meter pole. The perimeter of the lot directly in front of the pole is usually a point of average illumination when high-mast luminaires are use. Find the light level at this point by using the formula

$$E = \frac{N \times \text{LL} \times \text{CU} \times \text{LLF}}{\text{area}}$$

where E = level of illumination
N = number of luminaires
LL = initial lamp lumens
CU = coefficient of utilization
LLF = light-loss factor

The number of luminaires and the lamp lumens are known. The CU can be read from the photometric data given earlier in the chapter in Fig. 12-5. Since the distance from the pole base to the perimeter is 76.2 meters (see Fig. 12-9), with the 39.5-meter pole the transverse ratio is 76.2/39.5 = 1.93. This yields a CU of 0.28 in each direction, or a total CU of approximately 0.56.

For the 60-meter pole, the ratio is 76.2/60 = 1.27. Reading from the utilization curve, the CU is 0.22 × 2 or 0.44.

The light-loss factor was determined earlier to be 0.9 × 0.95 = 0.855.

Substituting in the equation for the 39.5-meter pole,

$$E = \frac{12 \times 140{,}000 \times 0.56 \times 0.855}{152.4 \times 152.4} = \frac{804{,}384}{23{,}226} = 34.6 \text{ lux}$$

This is the illumination level at the perimeter point. Given a 5.38-lux maintained light level, the average-to-minimum uniformity ratio is 34.6:5.38, or about 6.4:1. This is higher than necessary, but acceptable, because the main consideration in this case is to have enough illumination for the television camera to operate.

Substituting in the equation for the 60-meter pole,

$$E = \frac{6 \times 140{,}000 \times 0.44 \times 0.855}{152.4 \times 152.4} = \frac{316{,}008}{23{,}226} = 13.6 \text{ lux}$$

This system provides average-to-minimum uniformity of 2.5:1, since the average level is about 13.6 lux and the minimum 5.38 lux.

Step 11. Decide which system to use. The reduction in luminaire quantity from 12 to 6 will not begin to pay for the added cost of the higher pole. However, there will be a saving in power costs. Since power costs vary from place to place, the only way to make a choice between the 39.5-meter pole, the 60-meter pole, or something in between is to analyze local operating costs and balance these with initial costs. Another factor to consider is whether there are any restrictions on pole height in the area.

Step 13. Since the area in the example is protected by a television surveillance system, Chap. 13 should be checked for other factors that must be considered before the lighting system is installed.

13

Security Lighting Systems

Because of vandalism and theft, many industrial areas are protected by some sort of security system—either personnel on guard or a combination of a guard system and a closed-circuit television surveillance system. Security systems can be used to observe critical areas, to check on areas where alarms have been sounded, or to identify personnel entering sensitive sectors.

Television surveillance systems are effective both indoors and out, and should be designed in conjunction with the lighting system. General design procedures are the same as those used in other forms of indoor or outdoor area lighting. However, when security lighting systems are designed, it is always important that there be no dark areas or heavy shadows, and no bright light source that interferes with the view from the guard booth or is in direct line with television cameras.

Fence lighting and building floodlighting can be part of a security system. Their design is covered in this chapter.

Emergency lighting is intended more for safety than for security and will be covered in Chap. 14.

Visual Performance

Before designing a security lighting system the designer should understand the performance of the eye. Experiments have shown that a guard carrying a flashlight or lantern is not able to detect the presence of intruders as well as a guard whose eyes have become adapted to a fairly low light level. If the general surround is lighted to a low level (perhaps 11 lux, or 1 footcandle), then the guard will become adapted to the patch luminance of the flashlight—which could be 100 lux or more. In this case, the guard could walk within 10 meters (33 feet) of a darkly clad intruder and not see the intruder. In the same way, a searchlight moving around an area prevents a guard from seeing low-contrast objects outside the beam, particularly if they are not moving.

The whole security system should be designed as if the guard or television camera were the audience at a play and the intruder were the actor. Everything should be done to ensure that the audience has a good view of the actors and that there are no extremely bright spots in the immediate surround which would alter the audience's adaptation to darker areas.

Conversely, the approaching intruder should not be able to see the audience (the guards). The intruder can be faced with a barrage of lights, which will prevent dark-adapted eyes from seeing beyond them. This makes the intruder realize that unseen observors may be present. The resulting fear can increase adrenalin flow, which increases pupil dilation, which again increases the effect of glare.

If planning a glaring system, however, make sure that it will not interfere with workers entering a plant or with any neighbors in the vicinity.

Gatehouse Lighting

If the guard or guards are using a sentry box, or gatehouse, care should be taken that they can see out but intruders cannot readily see in. One way of ensuring this is to mount luminaires on the roof of the gatehouse itself. Also, any lighting necessary inside the gatehouse can be on a dimmer or screened from the guards, so that they maintain their adaptation to the dark.

Glossary of Terms

At the end of this chapter there is a glossary of technical terms used in television. Since television surveillance systems are a combination of lighting and television, it is helpful to be acquainted with the language used in both areas.

Components of TV Surveillance Systems

Components of a closed-circuit television surveillance system include the following:

1. The television camera
2. The transmission medium, such as coaxial cable
3. The television monitor
4. The power supply
5. One or more monitor observers
6. Accessories, such as pan and tilt mechanisms and remote lens-selection and adjustment mechanisms

Television systems can also be equipped with motion detectors that sound an alarm if any change occurs in the picture.

A closed-circuit system is similar to a standard television broadcast system, except that the picture information is transmitted to the screen by means of a coaxial cable instead of through space.

The camera lens focuses the scene to be televised onto the vidicon or other light-sensing element of the camera. Different kinds of photoconductive layers can be used on a vidicon's faceplate; some are more light-sensitive than others. The image on the faceplate is scanned by an electron beam. The amount of light that the beam encounters at any instant determines the camera-tube current, so that an output voltage proportional to the light intensity is obtained. The output voltage generated in this way is called the *video signal*.

The video signal is amplified and transmitted to the monitor, where it is converted to visual form. Pulses from a synchronizing generator lock together the scanning beams in the camera and the monitor.

Lenses

The lens for a camera can be chosen on the basis of its transmission efficiency, focal length, and lens-speed designation. Transmission efficiencies are usually between 70 and 90 percent; this figure can be obtained from manufacturers.

FOCAL LENGTH

Focal length is the distance between the sensing element of the camera and the effective center of the lens when the lens is focused on infinity. If an object at infinity is focused at a point 50.8 millimeters (2 inches) behind the lens, then the focal length for that lens is 50.8 millimeters (2 inches).

The greater the focal length, the larger the image of the object being viewed; but the width of field at a given distance is narrower with an increased focal length. Table 13-1 shows the relationship between lens size, focal length, and horizontal and vertical field of view. This information can be used to obtain either the width of field (W) or the lens-to-subject distance (D) in a particular application. The formula to use is

$$\frac{\text{Focal length}}{\text{Scanned-area width}} = \frac{D}{W}$$

This formula can be used with millimeters and meters or inches and feet. The scanned-area width depends on what is known as the *aspect ratio*.

Table 13-1 Relationship between lens size, focal length, and field of view

Lens, millimeters	Focal Length		Horizontal Field of View, degrees	Vertical Field of View, degrees
	Centimeters	Inches		
6.5	0.66	0.26	88	65
12.5	1.27	0.5	53	39
25	2.54	1.0	28	21
50	5.0	2.0	14	10.5
75	7.6	3.0	9.6	7.2
100	10.2	4.0	7.1	5.3
150	15.2	6.0	4.8	3.6
300	30.5	12.0	2.2	1.7

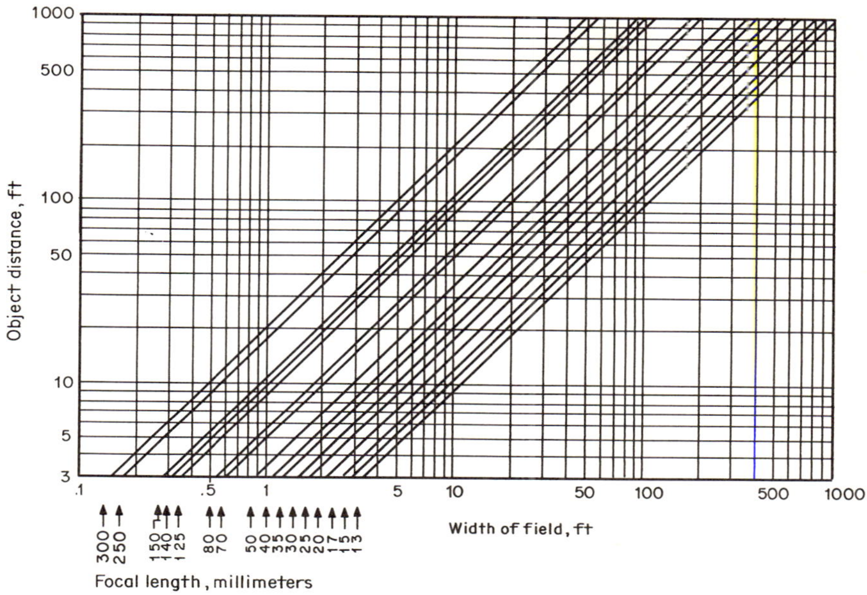

Fig. 13-1 Chart which can be used to find the width of the area covered by a particular lens at a specific distance. It can also be used to determine focal length or the correct object distance. (It is reprinted with permission from the General Electric Company, Tube Products Department.)

This is the ratio or proportion of the picture width to the picture height. It is 4:3 in present television systems, and so the scanned area of the light-sensitive element of a vidicon is 9.5 millimeters (3/8 inch) high and 12.7 millimeters (1/2 inch) wide, to correspond to this ratio.

The scanned-area width is thus 12.7 millimeters (1/2 inch). The formula to determine horizontal width of field can be rewritten

$$W = \frac{12.7 \times D}{\text{focal length}}$$

To determine focal length, use

$$\text{Focal length} = \frac{12.7 \times D}{W}$$

The vertical field of view (V) is three-fourths of the width of field, and can be obtained by this formula: $V = 0.75 \times W$. This is the height of the area covered.

Lens Selection Guide The same information can be obtained by using the lens selection guide in Fig. 13-1. Suppose that there is a particularly critical area where the television camera must be able to show clearly a 1.8-meter (6-foot) man. This is the height of the image that would fill the light-sensing element. Since $V = 0.75\ W$, if $V = 1.8$ meters, then $W = 2.4$ meters or 8 feet. Assume that the best position for camera placement is 21 meters (70 feet) from the area to be viewed. To determine the correct focal length to use, refer to Fig. 13-1 and follow these steps:

1. Find the width of field (1.8 meters, or 8 feet) on the horizontal scale at the top or bottom of the graph (Fig. 13-1).

2. Find the object distance (21 meters, or 70 feet) on the vertical scale at the left or right of the graph.

3. Follow these two lines to their point of intersection, and go down the nearest diagonal line to read the focal length in millimeters at the foot of the graph. In this case, the 8-foot and 70-foot lines intersect at the diagonal line for a 125-millimeter focal length.

The guide can also be used to determine the width of field if the object distance and focal length are known, or the object distance if the focal length and width of field are known.

Zoom Lenses A *zoom lens* is a special camera lens whose focal length or angle of view is continuously adjustable and always in focus on an object or scene within its range. Zoom lenses are often used in surveillance work, because they can zoom in on anything suspicious and bring it into view, while normally a wider area can be covered with the same camera.

LENS SPEED

The ratio of the focal length of a lens to the effective diameter of the maximum lens opening is the *f-stop number* of the lens. This number indicates the *speed* of the lens, or the amount of light it can pass to the light-sensitive element in the camera.

More light will be passed by a lens when the effective lens diameter is increased and the lens designation is decreased; that is, as the f-stop number decreases, the aperture size, lens speed, and light transmission all increase. Opening the lens by one f-stop increases the light transmitted by one-half. For security lighting, lenses with low f-stop numbers are generally used, because light levels are low.

Depth of field is the distance between the nearest and furthest points from the camera that appear to be in focus at a given setting for a particular camera. Depth of field is affected by the aperture, or f-stop, of the lens, as well as by the focal length of the lens and the amount of light available on the scene. Greater depth of field can be acquired by making the aperture size smaller, which means making the f-stop number larger. Manufacturers can supply tables giving depth-of-field limits for representative f-stops at various object distances.

The light available on the scene is directly related to the depth of field. If less light than needed is entering the camera, opening the aperture improves the picture presentation but cuts down depth of field. If there is more light than needed, the aperture can be closed down, which increases depth of field.

Light-Sensing Element

The *vidicon* is the light-sensing element of a television camera. Differences are achieved by using different photoconductive layers on the vidicon's faceplate.

Cameras are generally grouped according to their light sensitivity. The *dynamic range* of a camera is the range of illumination within which it will operate effectively. Most cameras will transmit a clear video picture from full sunlight down to some minimum level of illumination. Thus the dynamic range of a camera is greater if it contains a more sensitive light-sensing element.

Daylight cameras are designed to operate only at high levels of illumination. Their use is restricted to areas which are well lighted, such as indoor areas, doorways, or front gates. Many daylight cameras will suffer target burn and be damaged if they are aimed toward the sun or focused on a high-contrast scene for a long period of time.

Low-light-level (LLL) cameras will transmit a clear video picture from sunlight conditions to as little illumination as 0.00054 lux (0.00005 footcandle). There are many variations in sensitivity within the low-light-level category, and advances in camera technology are continually being made. Manufacturers should always be consulted about camera choice.

Levels of Illumination

In most cases, an outdoor maintained level of illumination of 22 to 33 lux (2 to 3 footcandles) is adequate for security lighting. These levels can be less if there is very little lighting on adjoining property or nearby roads. In that case, the guards can adapt to lower light levels because there are no higher-brightness areas in the immediate surround.

When calculating light levels for security lighting, the designer must use the light-loss factor that applies to the end of the relamping period.

This is the time at which luminaires are cleaned and relamped. Security lighting levels should never fall below recommended minimums.

For television-surveillance lighting, the light level must never be less than that required to activate the particular television camera that is in use. What counts is how much light reaches the light-sensing element of the camera. This is known as *faceplate illumination*, and camera manufacturers publish data on how much faceplate illumination is needed in order for a particular camera to operate properly.

In order to determine the sensitivity needed in a light-sensing element, the designer must know both the scene illumination and the scene reflectance. The surroundings can absorb 80 percent or more of the scene illumination. Table 13-2 gives some reflectance values. These are wide approximations and should be adjusted to meet different situations.

Light is absorbed by the lens of the camera, particularly if more than one lens is used. The lens transmission efficiency as well as the f-stop that will be used must be obtained from the lens manufacturer.

When all this information has been gathered, the necessary faceplate illumination can be determined from the following formula, provided that the lens will be operated wide open:

$$E_f = \frac{E_s \times RT}{4(\text{f\#})^2}$$

where E_f = faceplate illumination
E_s = scene illumination (lux or footcandles)
R = scene reflectance, expressed as a decimal between 0 and 1
T = lens transmission efficiency, similarly expressed
f# = f-stop number of lens, or focal length divided by maximum lens opening

For a sample problem, assume that a parking lot is being equipped with a television monitoring system. A parking lot full of automobiles has a reflectance of 40 percent, but asphalt reflects only 5 percent of the light

Table 13-2 Approximate reflectance values

Scene Being Televised	Reflectance, Percent
Empty asphalt parking lot	5
Parkland area, trees and grass	20
Red brick building	35
Unpainted concrete building	40
Parking lot with automobiles	40
Smooth surface aluminum building	65
Glass windows and walls	70
Snowy field with chain-link fence	85

reaching it; a reflectance value of 20 percent would probably be a good compromise figure to use. The minimum specified light level in this case is 21.52 lux (2 footcandles), the lens transmission efficiency is 90 percent, and the f-stop number to be used is 1.4.

Substituting in the formula,

$$E_f = \frac{21.52 \times 0.2 \times 0.9}{4\,(1.4)^2} = \frac{3.9}{7.84} = 0.5 \text{ lux } (0.045 \text{ footcandle})$$

A faceplate illumination of 0.5 lux is therefore necessary in this case in order for the camera to be able to survey the darkest sections. This tells the designer the sensitivity of the light-sensing element which would be required in this particular application.

Coordinating the Camera Choice with the Lighting

Generally, cameras containing light-sensing elements with greater sensitivity will cost more than those which contain elements which are less sensitive. The sensitivity needed depends on the amount of light available; if the proposed camera will not operate effectively under the existing or proposed lighting system, the question arises whether to buy a more sensitive camera or to increase the level of illumination in the area. This decision depends on the particular situation.

If the camera and luminaires are already in place, the lighting system may need only cleaning and relamping to increase the illumination level. If minor changes will not improve the situation enough, then the designer must decide whether to replace the camera or to improve the lighting system. The same situation can arise when a new television surveillance system is to be installed. The designer must weigh various factors.

One consideration should be the size of the area and the age of the existing lighting system. If it is a small area with a system that has been in place for many years, then it would probably be advisable to replace the lighting system rather than buy a very sensitive camera. If the area is large and the lighting installation fairly new, then buying a camera to meet the existing light levels might be the best choice.

The type of lamp is also a consideration in camera choice. Some cameras are rated under a light source of a particular color and will not perform as well under another light source. Camera manufacturers should be consulted on this point, as well as on the lens speed which would be best for the situation.

Costs are a major factor, whether a new installation is being designed or an older one updated. The cost of cameras varies between different types, since light-sensing elements need to be replaced periodically.

Operating costs are another consideration. The added power cost incurred by increasing the level of illumination in an area by 10.76 lux (1 footcandle) is approximately 40 dollars per year per 10,000 square feet if high-pressure sodium (HPS) sources are used. This is based on 4000 hours of operation per year and a power rate of 3 cents per kilowatthour. Equipment costs are not significantly higher with increased light levels, because higher-wattage, more efficient sources can often be used. Added camera costs can be balanced by power savings, and vice versa.

Only when all these costs have been determined can the best compromise between the camera and lighting system be made.

Sources

High-pressure sodium lamps are the most efficient light sources for this application. Their color matches the response of most cameras.

Lighting System Design

Techniques for outdoor or indoor security lighting are primarily the same as for other applications, except that luminaires should always be mounted so that they do not shine directly into a guardpost or into television cameras. If a camera pans through a light source, the picture will momentarily be obscured, and in some cases the camera will be damaged. Directing light away from the camera may require more poles and/or more luminaire locations than would ordinarily be needed.

Earlier chapters should be consulted for correct lighting design techniques. Design should satisfy minimum levels of illumination, so that there are no dark areas. This may mean that spacing is less than the maximum recommended by the manufacturer of the equipment. The system must also be slightly overdesigned, so that it will operate effectively even if a lamp should fail. If light comes from more than one direction, there will be fewer shadows.

The proposed luminaire choice should have a restrike time of 1 minute or less following a momentary power outage. High-pressure sodium sources meet this requirement. Also, there should probably be some type of incandescent system, separately powered, that will be used while the lamps are warming up.

Out of doors, glare is reduced by cutoff luminaires which deliver no light at all at vertical angles above 90 degrees. (Some cutoff luminaires deliver some light above 90 degrees; check photometric data for information.) These cutoff luminaires are not distracting to the viewer of a television monitor and make it easier for the viewer to spot anything suspicious in the area being watched.

High-mast poles are a good choice for television surveillance systems for three reasons:

1. The luminaires are out of the way of the camera and are therefore not a distraction.

2. Shadows are cut down, since the taller the pole, the shorter the shadows.

3. Higher poles make it possible to light a greater area from one location, thus reducing pole costs and the number of obstructions in the area.

Before proposing a lighting system, the designer should obtain a layout of the types of cameras to be used and their locations, if this has been decided. If not, the lighting designer should consult with the camera manufacturer, as well as the persons planning the installation. Cameras can be fixed in place or can be mounted so that they can pan through a given area. In a large installation, more than one type of camera may be used. All these factors affect the final lighting system design.

Fence Lighting

High-pressure sodium luminaires are commonly used for fence lighting in conjunction with television surveillance systems. Either roadway luminaires or floodlights can be used, depending on the type of system desired.

Roadway luminaires are used when a roadway parallels the fence that is being lighted. Generally, in this type of lighting it is expected that the luminaire will cover 4.6 meters (15 feet) on either side of a fence, as well as the fence itself. The luminaire should never overhang the fence, since this would provide an arm over which an intruder could throw a rope.

A typical fence-lighting system consists of 9.1-meter (30-foot) poles on 21.3-meter (70-foot) centers using 400-watt HPS cutoff luminaires to cover an area 9.1 meters (30 feet) wide. This gives a minimum level of illumination of 32 lux (3 footcandles).

Spacing depends on the desired minimum level of illumination and the planned uniformity. If there is not much choice in spacing, then different levels of illumination can be obtained by using sources of higher or lower wattage. If cutoff luminaires are used, spacing must be slightly less for the same results.

Floodlights mounted on short poles can also be used for fence lighting, particularly for security systems. The floodlights produce a glare barrage in the direction of the fence, which makes it difficult for someone approaching the fence to see beyond the poles. This is the best system for surveillance lighting, since there is no direct luminaire brightness behind the pole.

Poles 4.6 meters (15 feet) high are commonly used. They are placed at a distance behind the fence so that the floodlight beam covers as wide a section of fence as possible. Poles can be placed as much as 21 meters (70 feet) behind the fence.

Building Floodlighting

Building floodlighting can be adequate for security, as well as aesthetically pleasing. With care in the design procedure, both objectives can be met.

Design techniques are the same as for area lighting, except that the mounting height for the floodlights is replaced by the setback of the floodlights from the building. Sometimes floodlights are pole mounted if very high structures are involved. These steps should be followed:

1. Determine the desired level of illumination from Chap. 1.

2. Decide on the setback of the floodlight. The top of the building is the most difficult part to light. With any one floodlight type, the highest light level that can be produced at the top of the building is produced when the floodlight setback is approximately three fourths of the vertical distance, or height of the building. When floodlights are closer to the building than half the building height, the lighting will have a grazing effect that emphasizes any architectural detail or texture in the surface. Locating floodlights farther away than three fourths of the building height will only result in a loss of efficiency.

A setback of three fourths of the building height will produce the most uniform appearance on the building. Any time the light level drops evenly from bottom to top and is free of bright spots or streaks, the building will appear to be uniformly lighted. If the surface of the building is fairly flat with subdued architectural detail, the light levels should be close to a maximum-to-minimum ratio of 3:1. For structures with uneven surfaces, the light and shadow produced by the light striking the surface will mask all but major changes in light level (that is, all but those over 10:1). The level of illumination can change from bottom to top by a factor of 20:1 for very tall buildings with uneven surfaces.

3. Select floodlight spacing. Spacing should be determine by the shape of the beam pattern of the floodlight being used. For wide-beam floodlights mounted one per location, the separation should not exceed twice the setback from the building. Further separation is possible if two floodlights are mounted in each location. If necessary, precise spacing can be determined by using isoillumination diagrams, as explained in Chaps. 9 and 10.

4. Determine the number of floodlights with this formula:

$$N = \frac{\text{total length of building}}{\text{spacing}}$$

5. Adjust the aiming for the best visual effect. For pole-mounted floodlights that are half the building height or more, the floodlights can be aimed midway up the structure. If a floodlight is ground-mounted and the setback is equal to at least half the building height, the aiming point should be at least two thirds of the way up the building. As the setback is reduced, the aiming point must go higher on the building in order to maintain the same uniformity. Field adjustments are almost always necessary in this type of lighting.

Glossary of Technical Terms for Television[1]

Aperture: An opening that limits the amount of light entering an optical system.
Aspect Ratio: The numerical relation or proportion of the picture width to the picture height. It is 4:3 in present television systems.
Bandwidth: The range of frequencies required to convey the visual or aural information being transmitted. The bandwidth of a television channel in the United States is 6 megacycles.
CCTV: Abbreviation for "closed-circuit television."
Channel: A specific band of frequencies assigned to each radio or television station.
Control Console: A piece of equipment that incorporates monitors for viewing separate images picked up by various television cameras in a system in addition to the switching and other control devices required.
Definition: The appearance of sharpness of an image when it is in focus.
Depth of Field: The latitude or range within which a lens will keep an object in sharp focus at a given setting.
Dynamic Range: A measure of the operating range of a pickup tube from low brightness to maximum brightness.
Field: Either of the two groups of scanning lines that make up the complete television frame in a two-to-one interlaced system.
Field Frequency: The number of fields scanned per second. In the United States the standard is 60 fields per second.

[1]Reprinted with permission from *Glossary of Technical Terms for Television*, EBI-3795, General Electric Company, Electronic Components Business Division, Tube Products Department, Owensboro, Ky, 1972.

Focal Length: The distance from the center of a lens to the surface on which the image is focused.

Frame. The complete television image containing (in the United States system) 525 lines or two interlaced fields.

Frame Frequency: The number of complete pictures or frames scanned per second.

F-Stop Number: The speed of a lens, or its ability to pass light. It is calculated by dividing the focal length of the lens by its diameter.

Image Orthicon (IO): A highly sensitive television camera tube.

Iris: An adjustable aperture built in to each camera lens to permit control of the amount of light passing through the lens.

Lens Speed: The ability of a lens to pass light. The larger the f-stop number, the slower the lens.

Lens Turret: An accessory on a television camera which holds several different lenses in an arrangement designed to permit rapid interchange through rotation.

Monitor: A high-definition television viewer connected directly to the camera output. In some closed-circuit applications, monitors are used where high definition in the reproduced image is necessary.

Pan: Achieving a panoramic effect by pivoting the television camera from left to right or from right to left across a scene.

Photoconductive: Having the property of changing resistance to electrical flow under varying degrees of illumination.

Remote-Controlled Camera: A television camera equipped with accessories to permit its manipulation and adjustment by means of electronic controls.

Resolution: The ability of a television system to distinguish and reproduce fine detail in the subject picked up by the camera.

Speed: The f-value or aperture ratio of a lens. Specifically, speed is the ratio of the focal length to the physical diameter of the lens.

T-Stop: A system of calibration for rating the speed of lenses. This direct system is based on actual light transmission.

Vidicon: A television camera tube of much smaller physical size then the image orthicon.

Zoom Lens: A special camera lens whose focal length or angle of view is continuously adjustable and always in focus on an object or scene within its wide range.

Bibliography

American National Standard Practice for Protective Lighting, RP-10, A85.1-1956 (R1970), Illuminating Engineering Society of North America, New York, 1970.

Baker, J. E., and S. L. Lyons: "Lighting for the Security of Premises," *Lighting Research and Technology*, vol. 10, no. 1, 1978, p. 10.

Cochran, Howard A.: "Lighting for Closed-circuit TV Surveillance," *Lighting Design and Application*, March 1974, p. 47.

Crowther, Charles: "Better Cameras or More Lights?" *Security World*, November 1976, p. 32.

Fundamentals of Television and Essentials of Optics, EBI-3711, General Electric Company, Electronics Park, Syracuse, N.Y., 1972.

Glossary of Technical Terms for Television, EBI-3795, General Electric Company, Electronic Components Business Division, Owensboro, Ky.

Prell, J. A.: *Basic Considerations for Assembling a Closed-circuit Television System*, NUREG-0178, Office of Standards Development, U.S. Nuclear Regulatory Commision, May 1977.

Rowe, Gordon D., and R. L. Paugh, "A Crime Time Show," *Light*, vol. 38, no. 3, pp. 19–23.

14

Emergency Lighting Systems

Emergency lighting is designed to protect life and property in the event of a failure in the normal power supply. It is mainly for the purpose of supplying enough light for persons to leave a building safely. In industrial situations, it is also necessary for workers to be able to see well enough to avoid contact with dangerous machinery.

Auxiliary or standby lighting is different from emergency lighting; it is intended to provide enough light for normal operations to continue when the power fails, or to provide light while the normal lighting system warms up after a momentary power interruption. Auxiliary lighting can be fixed or mobile. In larger installations, it is operated by a generator that may be used to supply other power needs as well as lighting.

Emergency lighting is permanently installed, and is often battery-operated. It, too, can sometimes satisfy other power needs, but its main purpose is to enable persons to exit safely from a darkened building.

Codes

Emergency lighting is controlled by federal, state, and local regulations. Chief among these are the *Life Safety Code* and the *National Electrical Code* of the National Fire Protection Association. These codes are not binding in themselves but form a national consensus standard, which has been adopted almost everywhere across the country. The *Life Safety Code* sets forth how much illumination is required, where it is required, and for how long. The *National Electrical Code* sets the mechanical and electrical requirements of the emergency lighting system.

The Occupational Safety and Health Act sets no standards for the equipment for emergency lighting. Inspectors of the Occupational Safety and Health Administration (OSHA) insist, however, that equipment be listed by Underwriters Laboratories and installed in accordance with the *National Electrical Code* and the *Life Safety Code*. Another OSHA requirement is that "every exit shall be marked by a readily visible sign where the exit or way to reach it is not immediately visible." However, OSHA requires general emergency lighting only in pulp, paper, and paperboard mills.

The *National Electrical Code* states that all emergency lighting equipment must have rechargeable batteries and a means for recharging automatically after a power failure.

The *Life Safety Code* requires that there be emergency lighting on the floor along the route that a person would take in exiting from a building or structure. This includes angles and intersections of corridors and passageways, stairways, landings of stairs, and exit doors. The light level specified is 10.76 lux (1 footcandle) measured at the floor.

An emergency lighting system must come on automatically in the event of an interruption in the normal lighting, no matter how that interruption is caused. The system must be so arranged that failure of one unit will not leave any area in darkness. Exit signs and directional signs must either be part of an emergency lighting system or be provided with a power source of their own.

The *Life Safety Code* calls for the operation of the system for at least ½ hour, and for 1 hour in hospitals and institutions. However, 1½ hours is specified in the *National Electrical Code*, and this more rigid standard should be adhered to.

In certain areas of industrial facilities, the emergency lighting must continue to operate longer than these specified times. Longer times might be needed for security, for hazardous areas, or for continuity of critical operations. The Illuminating Engineering Society of North America (IES) recommends that emergency lighting continue for longer periods in "guard headquarters, gate houses, fire headquarters, fire pump rooms, first-aid areas, central control rooms, boiler operating areas, emergency generator rooms, telephone switchboard rooms, electrical switch gear rooms, and areas where hazardous processes must be controlled or shut down in orderly sequence." The IES also suggests that emergency lighting be provided in toilets and washrooms used by key personnel who will remain in the building. The society recommends that battery-powered lighting should be considered for the emergency-generator room, if there is one, so that there will be light for troubleshooting if the generator fails to start.

In addition to the requirements mentioned for emergency lighting, there is another factor recognized by the *Life Safety Code*. It is essential that emergency lighting not be glaring, distracting, or directed into a person's eyes when that person is trying to exit from a building. The lighting should be as close to normal as possible, since this helps to prevent panic.

Levels of Illumination

Areas which are extremely congested, or could be in an emergency, such as corridor intersections, tops of stairwells, etc., should be more brightly lighted than uncongested areas. An acceptable level would be 32 lux (3 footcandles). Certain apparatus and equipment, such as alarm boxes, fire-fighting equipment, dangerous machinery, etc., should also be highlighted.

The vertical surface surrounding an exit door should receive between 22 and 54 lux (2 to 5 footcandles) of incident illumination. Illumination around the main exitway should be 129 lux (12 footcandles), given a 50 percent reflectance of the walls.

Incident light of 54 lux (5 footcandles) is required on exit signs by some codes, but this is usually not sufficient. Although a slightly higher level would be better, exit signs should not be so bright as to distract or obstruct visibility.

Offices or other areas where persons are familiar with the environment and there is no hazardous machinery can be lighted with pools of light. Although 5.5 lux (0.5 footcandles) ambient is probably enough, 3 to 5 times that much illumination may be needed in order to define the pathways to exits. People tend to walk toward the area with the greatest illumination.

Types of Systems

There are many approaches to emergency lighting. Some of those approved by the *National Electrical Code* depend on power from off-premise sources and would be disrupted if there were a public-utility power failure. Other approved systems are (1) individual battery-powered lighting units, (2) centralized storage-battery–powered systems, and (3) mechanically driven generator-powered systems.

UNIT SYSTEMS

Unit systems are self-contained, complete packages which can be placed anywhere. Some are box-shaped, with one or more lamps on top, and others are shaped more like an ordinary luminaire. Each unit contains a battery and device to energize the system in the event of a power interruption. When individual lighting units are used, exit signs must also have individual power supplies.

Most package units use incandescent sealed-beam PAR (parabolic aluminized reflector) lamps. If the lamp is located so that its full light pattern can be used, then a 25-watt, 6-volt lamp can cover about 92 square meters (1000 square feet).

PAR lamps come in different wattages and different beam patterns. If the pattern is wide in the horizontal plane and narrow in the vertical plane, the lamp will be most effective in covering a large open area. PAR lamps with long, narrow beam spreads are better suited for lighting such areas as hallways or stairwells.

Another approach to individual unit systems is to use emergency battery and inverter units in existing ac fluorescent fixtures. The inverters change the dc output of the batteries to ac, so that these units can activate the luminaires in the event of a power failure. Usually only a portion of the luminaire is utilized for this purpose, and so the light level is lower than normal. Such systems have the advantage, however, of using the customary sources of light, which instill a feeling of security and lessen any tendency to panic.

These units can be mounted apart from, on top of, or within the luminaires. Some models are designed to be used in the ballast channels, but care must be taken here that batteries can withstand high temperatures. Other devices are designed to be installed within the luminaire but not in the ballast channel. They can be mounted between lamps or in line with them. Most of these systems can be either field-mounted or factory-installed.

All of these individual battery units must be inspected on a regular basis. Some have a combination test switch and pilot light. Others use other means of ensuring that the emergency lighting system is always ready for use. Even if the batteries used are maintenance-free, the rest of the unit must be serviced regularly.

To determine the battery size to suit the load, the designer must sometimes convert ampere-hours to watts of emergency lighting. This is done by using the formula

$$\text{Watts} = \frac{\text{volts} \times \text{ampere-hours}}{\text{time}}$$

Volts is the operating voltage of the dc lighting, ampere-hours is the manufacturer's rating on the battery for a 1½-hour rate, and the time is the 1½ hours required by the *National Electrical Code*.

CENTRALIZED BATTERY SYSTEM

The second type of approved emergency lighting system is the centralized battery system, with or without an inverter. When no inverter is used, the lamps connected to the system must be incandescent, since they will have to operate on dc power. When inverters are used to convert dc to ac power, central battery systems can be utilized to provide power to any type of luminaire—incandescent, fluorescent, or high-intensity discharge (HID).

These systems may be designed in several ways. The entire general-lighting layout can be wired to both a normal power supply and the emergency battery supply. When the regular system goes out, the emergency system can automatically take over and run the same luminaires as are normally used. Exit signs can also be tied in to the emergency lighting system. As an alternative, only a part of the normal lighting can be hooked in to the emergency power supply. In the event of a power failure, the general lighting is close to normal but provides a lower level of illumination. Luminaires used only for emergency lighting can be installed and wired in to the emergency power supply. These will come on automatically when necessary.

Centralized systems can be used to keep public address systems, alarm systems, and electric locks operating in the event of a power failure.

Sensing devices that activate centralized systems must not be located centrally but must be in the areas where the emergency lighting might be needed. Otherwise, if a branch-circuit fuse should blow and one area of a building be darkened, the emergency lighting might not come on.

MECHANICALLY DRIVEN GENERATOR SYSTEMS

The third source of emergency lighting is the mechanically driven generator station. This can be used to operate incandescent, fluorescent, or HID sources which are equipped with an instant-on device. The generator system can be running continuously so that the emergency lighting is always on; it can be running at idle, ready to be automatically connected to the emergency lighting; or it can be shut down normally and activated when the normal power supply is lost. In the last case, the generator needs some sort of fast-start device.

Choosing a System

Which type of emergency lighting system to use depends on the individual situation. In new construction, it will probably be less costly to use some kind of central system which has been planned in conjuction with the normal lighting or is a part of it. When central systems use all or part of a building's normal lighting, the only extra inspection required is the regular monthly check of the battery and central system. The luminaires themselves can be regularly serviced as a part of the lighting layout.

Individual battery units will probably be less costly to use in existing structures or in very small installations.

Exterior Lighting

When designing an emergency lighting system, the designer must consider exterior lighting, both for safe exit and for building security. Remote lamps can be used and powered from inside the building; but if all components are to be outside, they should be entirely enclosed to protect against vandalism. It is important to make sure that the lighting does not blind persons exiting from the building.

Special Situations

Any emergency lighting equipment which is to be used in corrosive atmospheres should be suitable for this use. It should be sealed and enclosed in inert materials. As much of the equipment as possible should be located outside of the area.

If emergency lighting is needed in an explosive area, explosion-proof equipment should be used. This equipment is labeled as suitable for such use.

Temperature is another factor to consider. Storage batteries lose some capacity in cold temperatures, but the loss is not significant until -15°C (0°F). If possible, mount the battery outside the cold area, and the luminaire inside. If this can't be done, then a battery suitable for use in extreme cold should be chosen.

Bibliography

Carson, James R.: "Self-contained Fluorescent Emergency Lighting Pack Comes in a New Package," *Lighting Design and Application*, April 1977, pp. 28–29.

Cusano, Arnold J.: "What's Available, "*Lighting Design and Application*, May 1974, pp. 6–8.

Finkelman, Jay J., and Howard Brandston: "Emergency Lighting Under Emergency Conditions—A Dim View," *Lighting Design and Application*, May 1976, pp. 45–50.

Hunter, Thomas C.: "Inverter Systems," *Lighting Design and Application*, May 1974, pp. 15–20.

IES Lighting Handbook, 5th ed., Illuminating Engineering Society of North America, New York, 1972, pp. 14-10 to 14-12.

Kilkelly, Brian H.: "A New Code—A New Ballgame," *Lighting Design and Application*, May 1974, pp. 4–5.

Life Safety Code, An American National Standard, National Fire Protection Association, Boston, 1970, pp. 48–52.

National Electrical Code, National Fire Protection Association, Boston, 1978, pp. 529–533.

Peterson, Warren: "First, Some Fundamentals, "*Lighting Design and Application*, May 1975, pp. 6–8.

Petersen, Warren: "The Unit Approach, "*Lighting Design and Application*, May 1974, pp. 9–14.

Petersen, Warren A.: "When Emergency Lighting Is Not Routine," *Lighting Design and Application*, April 1977, pp. 22–24.

Scharf, Beverly W.: "The Changing Market," *Lighting Design and Application*, May 1974, pp. 21–22.

Silvers, J. J.: "Visual Aspects of Life Safety Lighting, "*Lighting Design and Application*, October 1972, pp. 17–25.

"What's New in AC Systems," *Lighting Design and Application*, May 1975, pp. 37–38.

15

Pole Design and Selection

Poles come in many shapes and materials, and their heights range from less than 3 meters (10 feet) to 46 meters (150 feet). There are so many variables involved in making the correct pole choice that the designer is cautioned to seek as much information as possible before making a final selection.

Design information is available from pole manufacturers. Luminaire manufacturers can also recommend poles that are suitable for use with their individual products.

This chapter includes basic background information which can be helpful in pole selection. Methods for mounting luminaires are given, and pole strength is discussed.

Glossary

A glossary of terms connected with lighting structures is given at the end of this chapter. The designer should become familiar with this specialized vocabulary.

Pole Style and Dimensions

Poles are made from a number of different materials, including seamless and welded aluminum; prime-painted, galvanized, or weathering steel; concrete; and solid or laminated wood. Prime-painted and galvanized steel poles up to 12.2 meters (40 feet) are available in straight-sided or tapered square cross section.

Poles less than 12 meters are shipped in one piece. Longer poles may be shipped in two pieces. Some shorter poles are hinged for ease in maintenance. Others have steps on the sides.

Poles are considered to be high-mast poles when luminaires cannot be conveniently serviced from the ground. Generally the height of high-mast poles is assumed to be 18.3 meters (60 feet). Lowering devices are used to lower luminaires for servicing.

Luminaire Mounting

Design procedures in earlier chapters should be followed to determine the number of luminaires needed in each location and their mounting height. Then a choice of pole and mounting method can be made.

POLE-TOP MOUNTING

Decorative post-top luminaires can be mounted directly on top of poles. Roadway luminaires are mounted sideways at the top of poles. Floodlights are normally mounted vertically or at some angle.

Fig. 15-1 A decorative post-top luminaire can be mounted with a tenon. On aluminum poles, a plate is also used. Simplex adapters are used to attach roadway luminaires to mounting arms. Davits are used to mount single roadway luminaires.

Post-Top Luminaires Most decorative luminaires can be mounted directly on 3- to 6-meter (10- to 20-foot) poles with a slip fitter which is built in to the luminaire. For 6- to 18-meter (20- to 60-foot) poles, a tenon such as that in Fig. 15-1 is used to mount the luminaire. The size of the tenon is either 5.8 centimeters (2⅜ inches) or 10 centimeters (4 inches), depending on the number of luminaires to be used and their weight and projected area. With aluminum poles, a square mounting plate (Fig. 15-1) can also be used for mounting two to four luminaires.

Roadway Luminaires Roadway luminaires are mounted on poles from 3 to 18 meters (10 to 60 feet) tall. Roadway-type luminaires are usually mounted on 1.2-meter (4-foot) or longer arms attached to the side of the pole by bolts or by simplex adapters such as that shown in Fig. 15-1. One to four adapters are welded to the top of the pole to accommodate the number of luminaires. Poles for roadway-type luminaires are supplied with the necessary bracket arms, so the arm length should be specified with the pole order.

Single roadway luminaires can also be mounted on davit arms, such as that in Fig. 15-1. These are placed on top of the pole and the luminaire placed on the end.

Floodlights Floodlights can be mounted with a trunnion, which is a flat steel strap with a single bolt hole to fasten the floodlight to the crossarm. Floodlights can be mounted directly on a flat plate or on a tenon with an adapter. Some floodlights may be equipped with a tenon adapter when they come from the manufacturer. A crossarm adapter may be used for mounting floodlights. This makes it easier to aim them or to mount them at an angle.

Figure 15-2 shows two adapters used for floodlights; one for use on a pipe, tenon, or pole top and the other for use on a crossarm or flat surface.

When ballasts are separate from floodlights, the floodlights require other mounting arrangements. Manufacturers should be consulted.

MOUNTING ON HIGH-MAST POLES

High-mast luminaires should be mounted in such a way that they can be lowered with a lowering device for servicing. This requires that the luminaire weight be distributed evenly around the mounting ring. For this reason, luminaires are usually mounted in a circular pattern.

On high-mast poles, floodlights can be mounted on arms extending from a central ring, but the weight load must still be distributed evenly around the ring. This mounting method can restrict aiming somewhat, which is a consideration when the system is designed.

When there is no lowering device on a high-mast pole, floodlights can be mounted on platforms and fastened with crossarm adapters. Platforms are designed for a maximum number of floodlights. This number is dictated by the weight, size, projected area of the floodlights, and the strength of the pole. The designer should specify the number of floodlights and the mounting separation required between them to prevent interference when the floodlights are aimed.

ORDERING EQUIPMENT

When ordering adapters or any other pole accessories, the designer should make certain that all the parts will fit together when they arrive on

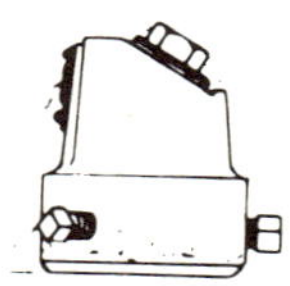

Pipe, tenon, and pole-top adapter

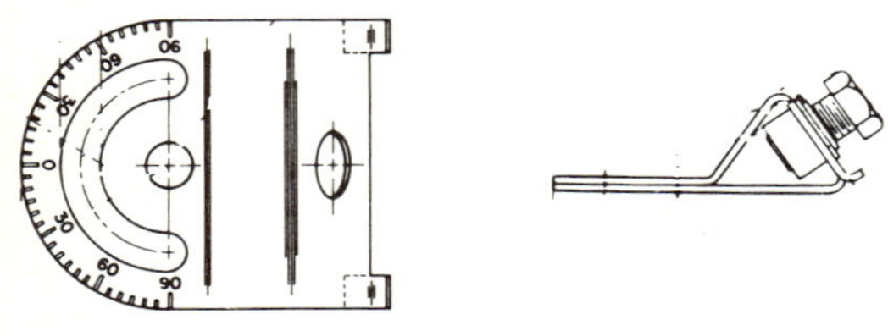

Crossarm or flat surface adapter

Fig. 15-2 Two types of adapters used to mount floodlights.

the site. Poles and luminaires are usually shipped from different locations, and so there is little opportunity to match parts beforehand.

Pole Strength

A lighting pole must support the weight of the equipment which will be mounted on it, and must withstand the effect of the maximum wind velocity to which it will be subjected. Therefore, a final pole choice cannot be made until these factors have been evaluated.

WIND VELOCITY

Wind velocity data has been collected for many years by the United States Weather Bureau. With this information, the United States can be divided into isotach zones, as in Fig. 15-3. Each zone on the map is coded for its maximum expected wind velocity. These are wind speeds measured 9.1 meters (30 feet) above the ground.

The map shown here is based on a 50-year mean. Other maps are published by the American Association of State Highway and Transportation Officials in *Specifications for Structural Supports for Highway Signs, Luminaires and Traffic Signals,* 1975.

The velocity values recorded on the map are isotach values. Winds may gust at higher velocities, so poles are designed to withstand wind velocities 30 percent higher than the isotach value for the area. Since measurements are made at 9.1 meters (30 feet), poles taller than 9.1 meters are strengthened still further.

There are some areas where unusual local wind conditions exist. These may necessitate the use of a greater isotach value than that shown on the map. If in doubt about the choice of a pole, consult the pole manufacturer.

WEIGHT AND EFFECTIVE PROJECTED AREA

The force of wind acting on an object is determined by multiplying the actual projected area of the object, its coefficient of drag, and the velocity pressure of the wind. The coefficient of drag takes into account the shape of the object; the shape can affect the wind force on the object by as much as 300 to 400 percent.

Multiplying the projected area of a luminaire by the coefficient of drag yields the *effective projected area,* or EPA, of that luminaire. Luminaire manufacturers publish EPA ratings for their products.

An EPA rating should use the highest drag coefficient as determined

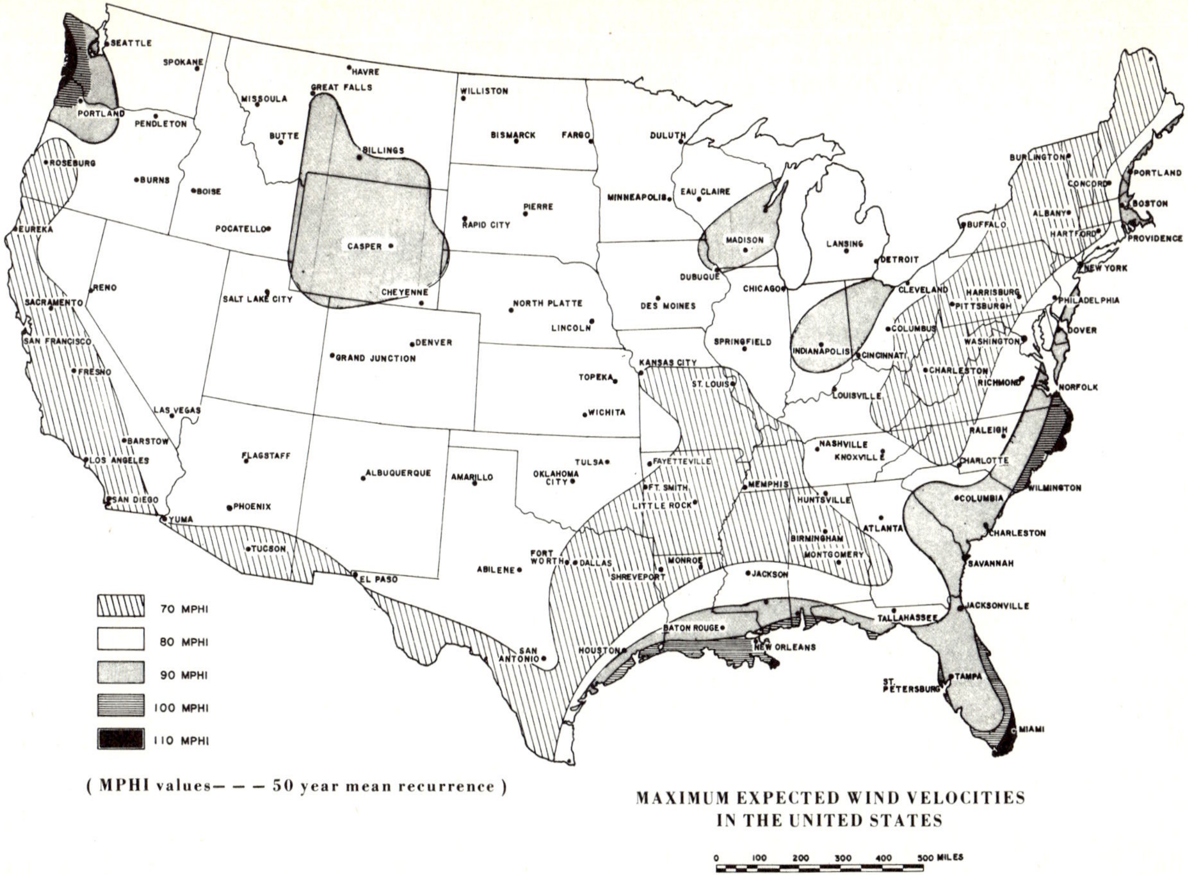

(MPHI values— — — 50 year mean recurrence)

MAXIMUM EXPECTED WIND VELOCITIES IN THE UNITED STATES

Fig. 15-3 It is important to know the maximum expected wind velocity in an area before choosing a pole. This information is collected by the United States Weather Bureau and compiled on isotach maps such as this one.

by orienting the luminaire in a number of positions in a wind-tunnel test. An EPA estimate based only on the projected area of the luminaire is not valid. There is also no reduction in the luminaire EPA when several luminaires are clustered on a pole, unless the EPA of the luminaire cluster has been tested and found to be lower.

The EPA of crossarms and any overhead wiring, banners, or signs must be added to the EPA for a luminaire. Weights of these various parts are also added to luminaire weight for total pole loading.

Luminaire and pole manufacturers publish data which gives the maximum wind rating for a particular combination of pole type, height, arm length, number of arms, weight, and EPA. The designer should select poles that have recommended loading values at least as great as the load to be applied. Any exceptions should be verified first by the pole manufacturer.

Pole Bases

Poles can be supplied with fixed or separate transformer bases. With a fixed transformer base the shaft is cut when the base is added. The pole height therefore remains the same. A separate transformer base can be used to increase mounting height. The base serves as a convenient wiring box.

An anchor-base pole should be wired at the time it is erected unless it is fitted with a handhole. Weatherproof switches and outlets can also be added to the base of the pole. An anchor-base pole is supplied with anchor bolts and a base template. These should be obtained beforehand so that the pole foundations can be poured and allowed to cure. This will prevent any delay in erecting the pole.

Foundations can be extended a few feet above the ground. This provides protection for the pole and is an inexpensive method of increasing mounting height. Qualified foundation designers should be consulted about pole foundations.

Checklist for Pole Choice

Given below is a check list to follow before making a pole choice. After this information has been gathered, the correct lighting pole can be determined with the aid of equipment manufacturers and catalog information.

1. CUSTOMER SPECIFICATION: If there is a specification, it is essential to know all of its details. It may call for special materials, special construction techniques, special strength requirements, etc.

2. POLE STYLE AND DIMENSIONS:
   - a. Material preferred, finish, and shape.
   - b. Mounting height. Nominal or exact?
   - c. Bracket-arm length.
   - d. Single, twin, or multiple bracket arms. If multiple, state number and alignment requirements.
   - e. Bracket-arm style preferred.
   - f. Transformer base, plain anchor base, or embedded base.
   - g. Pole-top dimension for crossarm or bullhorn.

3. LUMINAIRE TO BE SUPPORTED: Specific type and catalog number (if available).

4. OTHER EQUIPMENT TO BE SUPPORTED:
   - a. Projected area or dimensions of each item. Include alignment for determining direction of maximum wind force.
   - b. Some indication of shape of each item. Photograph, if possible.
   - c. Weight of each item.
   - d. Method of attachment to pole.
   - e. Manufacturer's catalog number (if available).

5. GEOGRAPHIC LOCATION: For determination of maximum wind velocity.

6. FOUNDATION IF EXISTING:
   - a. Anchor-bolt diameter.
   - b. Number of anchor bolts.
   - c. Bold circle diameter (usually center-to-center distance of diagonally opposite bolts).
   - d. Height of bolt projection from foundation.

7. MODIFICATIONS AND ACCESSORIES:
   - a. Handhole (height and radial position).
   - b. Electrical outlet (height and radial position).
   - c. Overhead wiring provisions.
   - d. Grounding connector.
   - e. Pole steps.
   - f. Other requirements.

Glossary of Pole Terms

Anchor Bolts: Bolts which anchor the pole base to the foundation in which they are encased. They are galvanized 30.5 centimeters (12 inches) on the top of the threaded end.

Back-up Bar: A steel bar which holds and releases handhole covers. The bolts fit through it and the cover.

Break-away Bases: Transformer bases of cast aluminum that will break at a given force. These are required on most highway jobs involving federal funds.
Bullhorn: An area lighting bracket that mounts over the pole tenon for use with floodlights.
Davit Poles: Lighting standards which are curved at the top. The luminaire is mounted to the pole parallel to the ground. They are used primarily for street lighting.
Double Bolt: The largest simplex, used with truss-type mast arms.
Embedded Pole: A lighting standard which is set directly into earth or concrete.
Festoon Box: A rectangular box in the side of the pole for electrical-outlet installation.
Ground Lug: An ordinary nut welded inside the base of the pole or inside the pole. A copper ground wire then grounds the nut.
Handholes: Holes in the side of the pole or base to allow wiring connections or initial installation and servicing thereafter.
Leaf or Nut Cover: A casting which covers the anchor-bolt nut.
Lowering Device: A device used for lowering the hinge poles for servicing. It is portable so that one unit can service several poles.
Mounting Height: The distance from the center of the luminaire to the bottom of the base.
Overhead Wiring: Electric wiring or cable that is strung overhead from pole to pole. Wiring enters the pole from the top.
Pole Steps: Steps consisting of rods placed at intervals on the pole to facilitate servicing.
Reinforcing Ring: An elliptical ring which welds into and around the handhole for reinforcement.
Scroll: A piece of light channel which mounts under the mast arm and back to the pole. It is ornamental only.
Self-leveling Nuts: Nuts on top of the anchor bolts that can be adjusted and leveled with one another so that the pole will be vertical.
Shaft Length: The distance from the top of the pole to the bottom of the base.
Shims: Pieces of steel that fit around the anchor bolts to level the base of the pole.
Shroud: The counterbalance piece of a hinge pole. It swings away from the pole when the luminaire is lowered. It also contains a locking device.
Simplex: Small steel castings used to attach the mast arm to the street lighting poles. There are two types, arm and pole simplexes and two sizes, one-bolt and two-bolt.
Slip Fitter: The straight end of the mast arm onto which the luminaire slips. The 5-centimeter (2.0-inch) size is primarily used; however, a 3.2-centimeter (1¼-inch) size is available.
Strain Pole: The end pole, when overhead wiring is used. It is usually of 7-gauge steel or heavier with 3.2-by 122-centimeter (1¼-by 48-inch) or 3.8-by 152.4-centimeter (1½-by 60-inch) anchor bolts.
Template: A cardboard print which provides the correct anchor-bolt placement.
Tenon: A 5- to 10-centimeter (2 to 4-inch) extension on area lighting poles. The luminaire ships over it and is attached with set screws.

Tie Backs: Two metal rods which extend from the mast arm back to the pole. They provide support for the arm.
Transformer Base: A steel or aluminum base to which the pole is mounted.
Truss: A type of mast arm which has a supporting member underneath. It is used almost exclusively on arms that extend beyond 2.4 meters (8 feet).
Up-and-Down Time: The period of time involved in erecting or dismantling a lighting standard completely.
Wire Entrance: The point at which the power cable enters the top or base of the pole.
Wire Inlet: An opening in the top or bottom of the pole, mast arm, or bracket for electrical wire or cable to enter.

Bibliography

Standard Specifications for Structural Supports for Highway Signs, Luminaires and Traffic Signals, American Association of State Highway and Transportation Officials, Washington, D.C., 1975.

Zeller, R. D.: "The Marriage of a Pole and Luminaire," *Lighting Design and Application,* June 1972, pp. 63–67.

16

Lighting Economics Recommendations

Over the life of a lamp as much as 90 percent of the cost of the light generated can be for power consumption. It is important, then, for a lighting designer to recommend the lighting system which not only provides a satisfactory installation but also uses energy most efficiently.

This chapter will explore various approaches to energy conservation, including the recommendations given in the joint publications of the American Society of Heating, Refrigerating, and Air-Conditioning Engineers (ASHRAE) and the Illuminating Engineering Society of North America (IES). A cost analysis sheet is included, as well as a chart that can be used to compare the efficacy of an existing installation with that of a possible replacement system.

ASHRAE Standards

The IES and ASHRAE have developed consensus standards which are being used by some state governments as the basis of building codes promulgated to save energy. The first of these, *Energy Conservation in New Building Design,* ASHRAE 90-75, was developed at the request of the National Conference of States on Building Codes and Standards.

The lighting section, Sec. 9 of ASHRAE 90-75, was formulated in cooperation with the IES and printed separately by the IES as *IES Recommended Lighting Power Budget Determination Procedures,* EMS-1.

The relevance of ASHRAE 90-75 to industrial lighting is limited. Its scope includes warehouse lighting, but only "those portion of factory and industrial occupancies which are used primarily for human occupancy, such as office space."

The ASHRAE-IES recommendations included in AHSRAE 90-75 apply to new building construction. The designer is expected to prepare a lighting power budget or limit for the entire structure—inside and out—before actually designing the lighting system. The procedures given in the standard are for preparing a power budget, not a lighting design. The budget applies to the entire building rather than to individual spaces. After compiling the power budget, the designer then designs the lighting system within or below that limit. This ensures efficient use of power.

ASHRAE and the IES have cosponsored other standards applying to existing facilities. The one that concerns industrial lighting is *Energy Conservation Existing Facilities—Industrial,* ASHRAE 100.4P, IES EMS 4.4, which was in draft form at the time of this writing.

This standard recommends that the designer follow the procedures given in ASHRAE 100.4P to compile a power budget for lighting in the industrial facility. Figures for actual lighting power consumption should then be gathered and the two sets of data compared. This study should make it obvious whether or not changing the existing lighting system will result in power savings. If the connected lighting power exceeds the lighting power budget, then the designer should select spaces considered most inefficient in lighting power utilization and modify them in order to reduce connected power.

The standard does not try to incorporate all the variables associated with modifying lighting systems for energy conservation. The designer is referred to the *IES Lighting Handbook* and qualified advisors for detailed information. Some specific recommendations are made, however, and the designer is encouraged to follow them.

Lighting Cost Comparison

The recommendations in the ASHRAE-IES standards are concerned with lamp and ballast watts—how much connected power the system will use. The lighting designer must also be concerned with other lighting expenses. The designer must calculate initial installation costs, maintenance costs, and other expenses in addition to power costs in order to make an educated choice of a lighting system.

Table 16-1 is a lighting cost comparison sheet. The designer gathers data on two or more systems and compares the costs of the systems over their total lives.

One of the systems being compared can be designated as the base, or the 100 percent point, for each element of cost. Each of the other systems is then compared with the base system. It is usually best to compare systems on the basis of some set level of illumination. This should be the illumination level needed in the particular application.

Before using this chart, the designer must gather much basic information, such as lamp and luminaire performance data and power and labor costs for the local area. Then pertinent data can be entered in the chart and valid comparisons made between different lighting systems.

This kind of comparison is only as accurate as the information that is used. The designer should try to get actual figures or figures gained by experience in similar situations. Rough estimates that do not take into account actual conditions should not be used.

Relighting

Often an industrial facility has a lighting system that has been in place for years and is using a disproportionate amount of connected power for the light that is being delivered. Completely replacing such a system with a

Table 16-1 Work sheet

	COST ANALYSIS / LIGHTING SYSTEM DESCRIPTION	BASE	II
BASIC DATA	1. Rated initial lamp lumens per luminaire		
	2. Rated lamp life (Hours) at ___ hours per start		
	3. Group replacement interval (Hours)		
	4. Average watts per lamp		
	5. Input watts per luminaire (Including ballast losses)		
	6. Coefficient of utilization		
	7. Ballast factor (Fluorescent)		
	8. Lamp depreciation factor		
	9. Dirt depreciation factor		
	10. Effective maintained lumens per luminaire (1 × 6 × 7 × 8 × 9)		
	10A. Average lux or footcandles on work surface 10 ÷ sq. m/luminaire (10 ÷ sq. ft./luminaire)		
	11. Relative number of luminaires needed for equal maintained lux or footcandles (10 of base system ÷ 10 of system compared)		
INITIAL COSTS	12. Net cost of one luminaire		
	13. Wiring and distribution system cost per luminaire		
	14. Installation labor cost per luminaire		
	15. Net initial lamp cost per luminaire		
	16. Total initial cost per luminaire (12 + 13 + 14 + 15)		
	17. Annual owning cost per luminaire (15% of 12 + 13 + 14)		
	18. Relative initial cost for equal maintained lux or footcandles (16 × 11 of system compared ÷ 16 of base system)		
OPERATING COSTS	19. Burning hours per year		
	20. Number of lamps group replaced per year (19 × lamps/unit ÷ 3)		
	21. Number of interim spot replacements (20 × Burnouts in GR interval)		
	21A. Number of lamps spot replaced per year—No group relamping (19 × lamps/unit ÷ 2)		
	22. Replacement lamp cost per year (20 *or* 21A × net lamp cost)		
	23. Labor cost for group replacements (20 × Group labor rate/lamp) at $___/lamp		
	24. Labor cost for spot replacements (21 × Spot labor rate/lamp) at $___/lamp		
	25. Cost of cleaning per luminaire per year		
	26. Annual energy cost per year (5 × 19 × ¢/KWH ÷ 100,000) at ______¢/KWH		
	27. Total annual operating cost per luminaire (22 + 23 + 24 + 25 + 26)		

	COST ANALYSIS	LIGHTING SYSTEM DESCRIPTION	BASE	II
TOTAL	28. Relative annual operating cost for equal maintained lux or footcandles (27 × 11 of system compared ÷ 27 of base system)			
	29. Total annual cost—owning and operating—per luminaire (17 + 27)			
	30. Relative total annual cost for equal maintained lux or footcandles (29 × 11 of system compared ÷ 29 of base system)			

newer, more efficient system will usually result in power savings as well as in increased light levels.

The chart in Fig. 16-1 can be used to find out if a new lighting system would have greater efficacy than an existing system. Before using the graph, the designer should obtain a plot plan of the entire industrial facility. This can be divided into smaller areas for comparison purposes. The designer should tour the facility and note the layout of the lighting, the number of luminaires, and what type they are. Supervisors should be asked to estimate the number of burning hours per year in each area.

The designer can then use Fig. 16-1 by following the steps given below.

STEP 1

For the area to be analyzed, determine the luminaire quantity, the area of the space, and the level of illumination provided by the existing lighting system. As an example, assume that forty 750-watt incandescent luminaires are used in an area of 1500 square meters to provide 160 lux.

STEP 2

Find the installed luminaire quantity (40) on the right-hand side of the lower section of the graph. Draw a line from this point horizontally to the diagonal line indicating the luminaire wattage of the lighting system presently installed (750).

STEP 3

From this point, draw a line straight up to the top of the lower section of the graph. This gives the connected load in kilowatts for the present system. In the example, the system uses about 30 kilowatts.

STEP 4

Continue going straight up to the diagonal line indicating the area covered by the existing system (in the example, 1500 square meters). The first number on each diagonal line is the area in square meters; the sec-

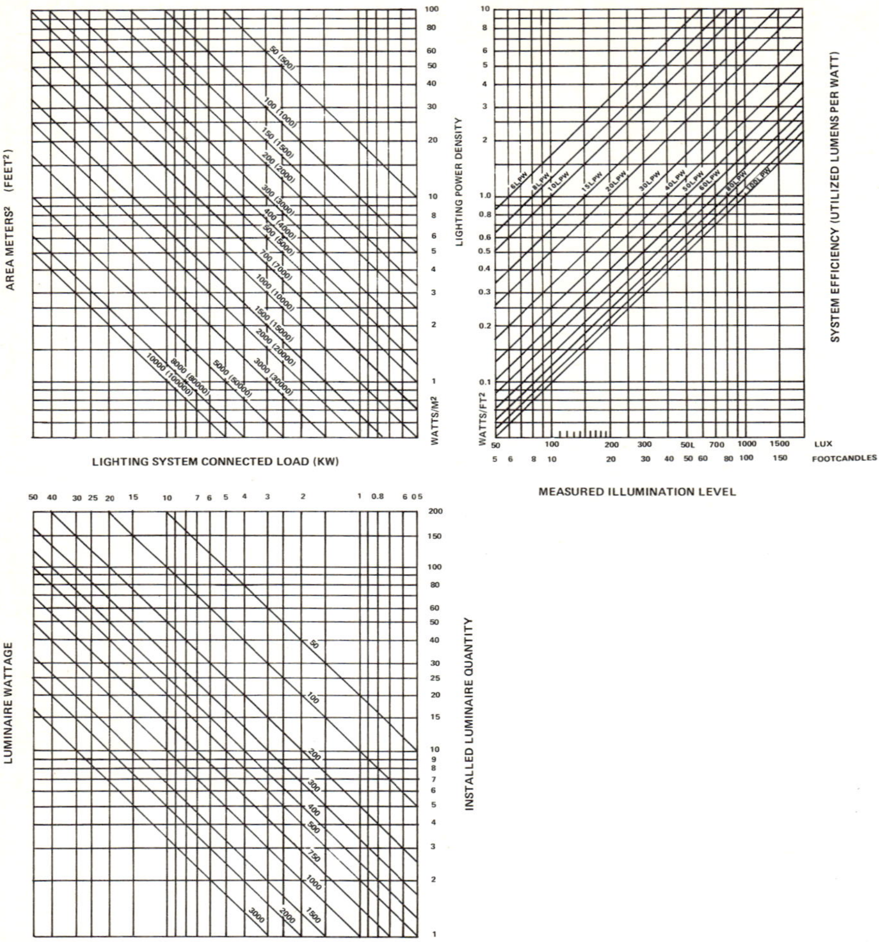

Fig. 16-1 Graph which can be used to compare the efficacy of an existing lighting system with that of a possible replacement system. The chart can be used in either lux and meters or footcandles and feet.

ond, in parentheses, is the area in square feet. The values are not equivalent; the chart must be used in one measurement system all the way through.

STEP 5

From the line indicating area, make a right-angle turn to the intersection marked *system watts/m²*. (If the chart is being followed in feet, find the ordinate marked *system watts/ft²*.) In the example, the system uses about 20 watts per square meter.

STEP 6

Continue in a straight horizontal line to the intersection with a line drawn up from the present level of illumination (160 lux). This gives the efficacy of the existing system, or the utilized lumens per watt. In the example, the efficacy is 8 lumens per watt, as noted on the diagonal line.

STEP 7

To compare the efficacy of the existing system with the efficacy of a different system, continue the horizontal line to the diagonal line marked with the lumens-per-watt rating of another type of luminaire. A 400-watt high-pressure sodium (HPS) system can deliver over 50 lumens per watt. From the diagonal line indicating 50 lumens per watt, draw a line down to the level of illumination. If the new system had as many 400-watt HPS luminaires as the present system has 750-watt incandescent luminaires—that is, 40—the light level would be increased to approximately 900 lux under the same power load; or the system could deliver 180 lux for about one-fifth of the power presently being used.

STEP 8

Use the information in Chaps. 4 and 5 to design a lighting system which uses a more efficient source. Often, fewer luminaires can be used to provide the same light level. Then follow the graph from the beginning again to compare the number of luminaires, connected load, watts per square meter, and efficacy with the figures for the existing system. There are considerable savings if all or part of the existing distribution system can be reused.

Repeat this procedure as many times as is desired before making a final lighting system choice.

Bibliography

Energy Conservation Existing Facilities–Industrial, ASHRAE Standard 100.4 P, IES EMS 4.4, second draft, American Society of Heating, Refrigerating, and Air-Conditioning Engineers, Inc., and Illuminating Engineering Society of North America, New York, 1978.

Energy Conservation in New Building Design, ASHRAE Standard 90-75, American Society of Heating, Refrigerating, and Air-Conditioning Engineers, Inc., New York, 1975.

IES Recommended Lighting Power Budget Determination Procedure, EMS-1, Illuminating Engineering Society of North America, New York, 1975.

Index